RETHINKING College Algebra

A Collaborative Approach to Mastery

First Edition

Mohammad Tabanjeh
Virginia State University

innovativeink PUBLISHING

Cover images Shutterstock.com

innovativeink PUBLISHING

www.innovativeinkpublishing.com
Send all inquiries to:
4050 Westmark Drive
Dubuque, IA 52004-1840

Print ISBN: 979-8-3197-3737-3
eBook ISBN: 979-8-3197-3738-0

Published in the United States of America

Table of Contents

Overview of the Textbook

This book is an introductory college algebra textbook designed to support active, small-group learning in the classroom. It presents essential algebraic concepts through a carefully structured sequence of seventeen concise chapters that integrate collaborative problem-solving with applied activities. The content emphasizes clarity and focus, allowing students to build foundational skills while engaging meaningfully with key algebraic ideas.

Core topics include linear and quadratic equations, inequalities, functions and their graphs, polynomial and rational functions, exponential and logarithmic functions, and systems of linear equations. Each lesson is intentionally designed to promote student participation, discussion, and peer-to-peer learning, shifting the classroom experience from passive lecture to active engagement.

A central feature of the text is its emphasis on collaborative learning. Students are encouraged to work in small groups to explore concepts, solve problems, and articulate their reasoning. This approach fosters deeper conceptual understanding, strengthens problem-solving skills, and supports improved retention. The structure aligns with current best practices in mathematics education, particularly in four-year colleges, community colleges, and co-requisite course settings.

The textbook also incorporates applied projects which connect algebraic concepts to real-world contexts. These projects are designed to reinforce learning, promote critical thinking, and provide opportunities for extended group investigation.

The streamlined organization of the material offers instructors flexibility in pacing and implementation. The text is well suited for introductory college algebra courses, developmental or transitional algebra programs, and co-requisite support courses. By combining essential algebra content with a practical, student-centered instructional design, this textbook provides an engaging and effective alternative to traditional lecture-driven materials.

Chapter 1

Linear Equations and Applications

A **linear equation** in one variable is an equation of the form $ax + b = c$, where a, b, and c are real numbers, and $a \neq 0$.
Sometimes, a linear equation is called a **first-degree equation**. A linear equation has exactly one solution. The solution is the numerical value of the variable. The graph of a linear equation is a line.
For example, $2x + 6 = 0$, $-y + 5 = 0$, $3p - 8 = 10$ are examples of linear equations in one variable.

Example 1: Solve $3x + 4 = 13$

Solution: The strategy is to isolate the variable x, so we subtract 4 from both sides of the equation.

$$3x + 4 - 4 = 13 - 4$$
$$3x = 9.$$

Divide both sides by 3, we obtain the solution

$$x = 3.$$

We can check our answer by substituting 3 in the original equation to make sure that 3 does, indeed, satisfy the equation.
left-hand side $= 3x + 4 = 3(3) + 4 = 13$ right-hand side $= 13$

Example 2: Solve $\frac{5}{6}x - \frac{4}{3} = \frac{3}{5}x + 1$

Solution: We first eliminate the fraction by multiplying both sides of the equation by the LCD of all fractions, which is 30.

$$(\frac{5}{6}x - \frac{4}{3})(30) = (\frac{3}{5}x + 1)(30)$$
$$25x - 40 = 18x + 30$$
$$7x = 70$$
$$x = 10.$$

Verify that $x = 10$ is a solution of the original equation.

Example 3: Solve $2 + 3x = -5 + 3x$

Solution: Combine the variable x, to do so subtract $3x$ from both sides of the equation.

$$2 = -5$$

which is impossible! It means there is no solution.

Example 4: Solve $p = 2l + 2w$ for l

Solution: Subtract $2w$ from both sides

$$p - 2w = 2l.$$

Divide both sides by 2

$$\frac{p - 2w}{2} = l.$$

Application of Linear Equations

Example 5: Two trains leave New York to Chicago. The first train travels at an average speed of 60 mph. The second train, which departs an hour later, travels at an average speed of 80 mph. How long will it take the second train to overtake the first train?

Solution: Since we interested in the time the second train travels, we let

$$t = \text{the number of hours the second train travels.}$$

Then

$$t + 1 = \text{the number of hours the first train travels,}$$

since the first train departs 1 hour earlier.
Using the equation **Distance=Rate** $\times$ **Time** or $\mathbf{d} = \mathbf{r} \times \mathbf{t}$.

	Rate	$\times$ Time	= Distance
First train:	60	$t + 1$	$60(t + 1)$
Second train:	80	t	$80t$

When the second train overtake the first train, they both have traveled the same distance. Therefor,

$$60(t + 1) = 80t$$
$$60t + 60 = 80t$$
$$60 = 20t$$
$$3 = t.$$

Thus, it takes the second train 3 hours to catch up with the first train.

Example 6: A total of \$20000 is to be invested, some in bonds and some in certificates of deposit (CDs). If the amount invested in bonds is to exceed that in CDs by 3000, how much will be invested in each type of investment?

Solution: Let x be the amount invested in CD, then $20000 - x$ is the amount invested in bonds.
Since the amount invested in bonds is to exceed that in CDs by 3000, then $20000 - x = x + 3000$, and if we solve this equation for x we have $2x = 17000$ and $x = 8500$. So 11500 invested in bound and 8500 invested in CD.

Example 7: Going into the final exam, which will count as two-third of the final grade, Mike has test scores of 86, 80, 84, and 90. What score does Mike need on the final exam in order to earn a **B**, which requires an average score of 80?

Solution: (exercise) ans: 78

Group Exercises 1

Linear Equations and Applications

(1) Solve each equation and check your answer. Write your answer in the exact fractional form.

(a) $-4(2x+1)-(x-2)=-11$

(b) $3(a+2)-2(a-3)=0$

(c) $7-(2x-1)=10$

(d) $\frac{2}{3}p=\frac{1}{2}p+\frac{1}{3}$

(e) $\frac{2y+5}{4}=\frac{3-4y}{5}$

(f) $\dfrac{x}{x-2}+3=\dfrac{2}{x-2}$

(g) $\dfrac{5}{2x-3}=\dfrac{3}{x+5}$

(2) Solve for the indicated variable.

(a) $ax-b=c$ for x

(b) $l-w=tw$ for w

(c) $y-y_1=m(x-x_1)$ for x

(3) Given the functions $f(x)=-2x-1$ and $g(x)=-\frac{6}{5}x+\frac{7}{5}$:

(a) Does $f(x)$ and $g(x)$ intersect?

(b) If no, explain why. If yes, at what point do they intersect?

(4) Ruby is driving along Interstate 20 traveling from Dallas to Marshall. She has her cruise control set to 65 miles per hour. As she passes a sign showing there are 50 miles to Marshall, she wonders how many minutes it will take her to get to Marshall if she maintains her present speed. (distance = rate × time) ans: 46.15 minutes.

(5) Suppose that you are trying to decide between two long distance services. Company **A** will allow you to call your family for \$0.15 per minute with a \$1 connection charge. Company **B** will charge you \$0.25 per minute, but there is no connection fee.

(a) Write the algebraic rule for the linear function $\mathbf{A(x)}$ that represents the cost of an x-minute phone call to your family using company **A**.

(b) Write the algebraic rule for the linear function $\mathbf{B(x)}$ that represents the cost of an x-minute phone call to your family using company **B**.

(c) If you talk to your family for 45 minutes, how much will company **A** charge you? How much will company **B** charge you?

(d) If you have \$5 to spend, how long will company **A** allow you to talk? How about company **B**?

(e) What is the length of the a phone call for which the companies charge the same amount?

Practice Exercises 1

Linear Equations and Applications

(1) Let $f(x) = \frac{4}{3}x + 8$ and $g(x) = -x - 5$, at what point do $f(x)$ and $g(x)$ intersect?

(2) Solve each equation. Write your answer in exact fractional form.

(a) $3 - 8u = -8u + 4$

(b) $3(2 - x) = 2x - 1$

(c) $\frac{3}{2}x = 2 - \frac{2}{3}x$

(d) $\frac{-4}{x+4} = \frac{-3}{x+6}$

(e) $\dfrac{4}{x-2} = \dfrac{-3}{x+5} + \dfrac{7}{(x+5)(x-2)}$

(3) Solve for the indicated variable.

(a) $2x - 5y = 7$ for y

(b) $8 - 2kx = -3x$ for x

(c) $A = \dfrac{1}{2}(a+b)h$ for b

(4) Suppose that you are going on a vacation first to Mexico, then Japan. You have \$5455.68 to spend on your trip. First you need to change your dollars to pesos so you can go to Mexico. The changer charges 7% and exchange rate is 1 dollar = 11.39 pesos. Round to the nearest hundredth after each conversion.

(a) After you change your money the first time, how many pesos do you have?

(b) You spend 21960.34 pesos in Mexico. Then you have to pay the changer again to convert to yen at a rate of 1 peso= 12.23 yen. How many yen do you have?

(c) In Japan you spend 110032.37 yen. Then you have to pay the changer again to convert your remaining money back to dollars. How many dollars do have left?

(5) Suppose you need to hire a painter to paint your house. Painter Jim charges \$20 an hour and \$50 for the material cost. Whereas painter Jill charges \$490 to paint the house.

(a) Write the equation that describes the price you will pay to hire each painter.

(b) Sketch the graph of both equations, then determine which painter you should use for any estimated number of hours of labor that the job will take?

Chapter 2

Quadratic Equations and Applications

A **quadratic equation** is an equation of the form $ax^2 + bx + c = 0$, where a, b and c are real numbers and $a \neq 0$. Sometimes the quadratic equation is called a **second-degree equation**.

Solving Quadratic Equations by Factoring

If $a.b = 0$, then either $a = 0$ or $b = 0$

Example 1: Solve $x^2 - 25 = 0$

Solution: $(x-5)(x+5) = 0$, then $x = 5$ or $x = -5$.

Example 2: Solve $x^2 + 4x = 0$

Solution: $x(x+4) = 0$, then either $x = 0$ or $x = -4$

Example 3: Solve $x^2 - 5x + 6 = 0$

Solution: $(x-3)(x-2) = 0$, then $x = 2$ or $x = 3$.

Example 4: Solve $2x^2 - 5x - 3 = 0$

Solution: $(2x+1)(x-3) = 0$, then $x = 3$ or $x = -\dfrac{1}{2}$

Solving Quadratic Equation by Completing the Square

Example 5: Solve $x^2 - 3x + 2 = 0$ by completing the square.

Solution: Rewrite the equation with the constant term on the right-hand side, $x^2 - 3x = -2$. Take $\frac{1}{2}$ of the coefficient of x and square it, $(\frac{-3}{2})^2 = \frac{9}{4}$.
Add the last number to both sides of the last equation to obtain $x^2 - 3x + \frac{9}{4} = -2 + \frac{9}{4}$.
Now, simplify and solve for x. $(x - \frac{3}{2})^2 = \frac{1}{4}$. Take the square root of both sides, $(x - \frac{3}{2}) = \pm\sqrt{\frac{1}{4}} = \pm\frac{1}{2}$, and add $\frac{3}{2}$ to both sides of the last equation. Therefore, $x = \frac{3}{2} \pm \frac{1}{2}$, i.e., $x = 1$ or $x = 2$.

Example 6: Solve $2x^2 - 4x - 2 = 0$ by completing the square.

Solution: Rewrite the equation with the constant term on the right-hand side, $2x^2 - 4x = 2$. Factor 2 out from the right-hand side, $2(x^2 - 2x) = 2$. Divide both sides of the last equation

by 2, $x^2 - 2x = 1$.
Now, find $\frac{1}{2}$ of the coefficient of x and square it, i.e., $(\frac{-2}{2})^2 = 1$, and add the number 1 to both sides of the last equation to obtain $x^2 - 2x + 1 = 1 + 1$, or $(x-1)^2 = 2$.
Solve the last equation for x, by taking the square root of both sides, $x - 1 = \pm\sqrt{2}$ which implies that $x = 1 \pm \sqrt{2}$.

Solving Quadratic Equation by The Quadratic Formula

The quadratic equation $ax^2 + bx + c = 0, \ \ a \neq 0$ has *two* roots often written in the form

$$\boxed{x = \frac{-b \pm \sqrt{b^2 - 4ac}}{2a}, \ \ a \neq 0}$$

Example 7: Solve $2x^2 - 3x - 3 = 0$ by the quadratic formula.

Solution: Since $a = 2$, $b = -3$ and $c = -3$,

$$x = \frac{-b \pm \sqrt{b^2 - 4ac}}{2a},$$

we have

$$x = \frac{-(-3) \pm \sqrt{(-3)^2 - 4(2)(-3)}}{2(2)}$$

$$x = \frac{3 \pm \sqrt{33}}{4}$$

Example 8: Solve $-5x^2 + 3x = 2$ by the quadratic formula.

Solution: We first rewrite the given equation as $-5x^2 + 3x - 2 = 0$, the $a = -5$, $b = 3$ and $c = -2$. Substituting into the quadratic formula,

$$x = \frac{-b \pm \sqrt{b^2 - 4ac}}{2a},$$

we have

$$x = \frac{-3 \pm \sqrt{3^2 - 4(-5)(-2)}}{2(-5)}$$

$$x = \frac{-3 \pm \sqrt{-31}}{-10}$$

$$x = \frac{-3 \pm \sqrt{31}i}{-10},$$

these are complex roots. Notice that $\sqrt{-1} = i$

The discriminant

The expression $b^2 - 4ac$ which can be found under the radical in the quadratic formula is called the *discriminant*. The quadratic equation $ax^2 + bx + c = 0$, $a \neq 0$ has exactly two roots, the nature of these roots are determined by the discriminant $b^2 - 4ac$.

(1) If $b^2 - 4ac$ is positive, then the quadratic equation has two different real roots.

(2) If $b^2 - 4ac$ is negative, then the quadratic equation has two complex roots (conjugate of each other).

(3) If $b^2 - 4ac = 0$, then $x = -\dfrac{b}{2a}$ is a repeated real root.

Example 9: Without Solving, determine the nature of the roots of the quadratic equation $3x^2 - 4x + 6 = 0$.

Solution: Since $a = 3$, $b = -4$ and $c = 6$, then

$$b^2 - 4ac = (-4)^2 - 4(3)(6) = 16 - 72 = -56.$$

Since the discriminant is negative, so the equation has two complex roots which are conjugate of each other.

Example 10: Without Solving, determine the nature of the roots of the quadratic equation $2x^2 - 7x = -1$.

Solution: First, write the equation in the standard form

$$2x^2 - 7x + 1 = 0,$$

then substitute $a = 2$, $b = -7$ and $c = 1$ in the discriminant.
Therefore,

$$b^2 - 4ac = (-7)^2 - 4(2)(1) = 49 - 8 = 41.$$

Since the discriminant is positive, thus the roots are distinct real roots.

Example 11: A fence is to be built to enclose a rectangular garden, if there is 64 feet of fencing and you want the area of the garden to be 256 square feet, determine the dimensions of the garden.

Solution: $2l + 2w = 64$ (perimeter) and $l.w = 256$ (area).
First, solve the first equation for either l or w, let us solve it for l.

$$l = \frac{64 - 2w}{2} = 32 - w.$$

Then substitute this value in the second equation $l.w = 256$ to obtain

$$(32 - w)(w) = 256.$$

Then solve the last equation for w.

$$\begin{aligned} 32w - w^2 &= 256 \\ w^2 - 32w + 256 &= 0 \\ (w-16)(w-16) &= 0 \\ w - 16 &= 0 \\ w &= 16 \end{aligned}$$

To find the value of l, we use the equation $l = 32 - w$. Then $l = 32 - w = 32 - 16 = 16$. Thus the dimensions are 16 feet long by 16 feet wide.

Group Exercises 2

Quadratic Equations and Applications

(1) Solve each equation by factoring:

(a) $x^2 - 9 = 0$

(b) $2x^2 - x = 0$

(c) $x^2 + x - 6 = 0$

(d) $2x^2 = 5x + 3$

(e) $4x^2 + 9 = 12x$

(2) Solve each equation by completing the square:

(a) $x^2 + 4x = 21$

(b) $x^2 - 6x = 13$

(c) $x^2 + 3x = 4$

(3) Find the real solutions, if any, of each equation using the quadratic formula:

(a) $x^2 - 4x + 2 = 0$

(b) $2x^2 - 5x + 3 = 0$

(c) $x^2 + 6x + 1 = 0$

(d) $4x^2 - x + 2 = 0$

(4) Without solving, determine the nature of the roots of each quadratic equation:
(a) $2x^2 + x + 5 = 0$

(b) $3x^2 + 2x = 5$

(c) $4v^2 = -4v - 1$

Problem I. The area of a rectangular window is to be 306 square centimeters. If the length exceeds the width by 1 centimeter, what are the dimensions?

Problem II. A ball is thrown vertically upward from the top of a building 96 feet tall with an initial velocity of 80 feet per second. The distance s (in feet) of the ball from the ground after t seconds is $s = -16t^2 + 80t + 96$.

(1) After how many seconds does the ball strike the ground?

(2) After how many seconds will the ball pass the top of the building on its way down?

Practice Exercises 2

Quadratic Equations and Applications

(1) Solve each equation by factoring:

(a) $x^2 - 25 = 0$

(b) $2v^2 - 50 = 0$

(c) $x^2 - 7x - 8 = 0$

(d) $3t^2 - 4t + 1 = 0$

(e) $6x^2 - x = 2$

(2) Solve each equation by completing the square:

(a) $t^2 - 2t = 15$

(b) $x^2 + 5x + 4 = 0$

(c) $2x^2 - 3x - 1 = 0$

(3) Find the real solutions, if any, of each equation using the quadratic formula:
(a) $4s^2 + s + 1 = 0$

(b) $x^2 - 4x - 1 = 0$

(c) $2x^2 + 4x - 3 = 0$

(d) $\frac{3}{4}x^2 - \frac{1}{4}x - \frac{1}{2} = 0$

(4) Without solving, determine the nature of the roots of each quadratic equation:

(a) $y^2 - 2y + 3 = 0$

(b) $-3z^2 + 2z + 5 = 0$

(c) $12s = 9s^2 + 4$

Problem I. A rock is thrown from the top of a cliff. The height of the rock, h, in feet, above the ground as a function of time, t, in seconds is given by:

$$h(t) = -16t^2 + 48t + 64 = -16(t - \frac{3}{2})^2 + 100$$

(1) How high is the cliff?

(2) What is the maximum height above the ground reached by the rock?

(3) How long after throwing the rock does it take to reach this maximum height?

(4) How long after throwing the rock does it take to reach a height of 32 feet?

(5) How long after throwing the rock does it take to hit the ground?

Problem II. An open box is to be constructed from a square piece of sheet metal by removing a square of side 1 foot from each corner and turning up the edges. If the box is to hold 4 cubic feet, what should be the dimensions of the sheet metal?

Chapter 3

Radical Equations and Factoring Methods

Certain types of equations can be transformed into quadratic equations that can be solved by the methods discussed in the previous chapters. One of the form that leads to either linear or quadratic equation is the **radical equation**. A radical equation is an equation in which the variable occurs in a radical, that is, a square root, a cube root, and so on.

Example 1: Find the real solution: $\sqrt{2t-7} = 3$

Solution: Since the equation contains a square root, square both sides,

$$(\sqrt{2t-7})^2 = 3^2$$
$$2t - 7 = 9$$

and solve for t.

$$2t = 16$$
$$t = 8$$

Check:
$t = 8$:
LHS: $\sqrt{2t-7} = \sqrt{2(8)-7} = \sqrt{9} = 3$ and RHS $= 3$, so since LHS=RHS, then $t = 8$ is a solution of the equation.

Example 2: Find the real solution: $\sqrt{13-x} = x - 1$

Solution: Since the equation contains a square root, square both sides,

$$(\sqrt{13-x})^2 = (x-1)^2$$
$$13 - x = x^2 - 2x + 1$$

and then put the equation in the standard form,

$$x^2 - x - 12 = 0$$

and solve for x by factoring.

$$(x-4)(x+3) = 0.$$

Then either $x = 4$ or $x = -3$.

Check:
$x = 4$: $\sqrt{13-x} = \sqrt{13-4} = \sqrt{9} = 3$ and $x - 1 = 4 - 1 = 3$
$x = -3$: $\sqrt{13-x} = \sqrt{13+3} = \sqrt{16} = 4$ and $x - 1 = -3 - 1 = -4$
So $x = 4$ is the only solution of the equation.

Example 3: Find the real solution: $\sqrt[3]{1-2x}-1=0$

Solution: The equation contains a radical whose index is 3. We isolate it on one side of the equation.

$$\sqrt[3]{1-2x}=1$$

Since the index of the radical is 3, we raise each side to the third power and solve for x.

$$(\sqrt[3]{1-2x})^3=1^3$$

Simplify

$$1-2x=1$$
$$2x=0$$
$$x=0$$

Check: $\sqrt[3]{1-2x}-1=\sqrt[3]{1-2(0)}-1=0$. Therefore, the solution set is $\{0\}$.

Example 4: Find the real solution: $\sqrt{2x+3}-\sqrt{x+1}=1$

Solution: First, we choose to isolate the more complicated radical expression (in this case $\sqrt{2x+3}$) on one side of the equation.

$$\sqrt{2x+3}=\sqrt{x+1}+1$$

Now square both sides.

$$(\sqrt{2x+3})^2=(\sqrt{x+1}+1)^2$$
$$2x+3=(\sqrt{x+1})^2+2\sqrt{x+1}+1$$
$$2x+3=x+1+2\sqrt{x+1}+1$$
$$2x+3=x+2+2\sqrt{x+1}$$

Because the equation still contains a radical, we isolate the remaining radical on the right side and again square both sides.

$$2x+3-x-2=2\sqrt{x+1}$$
$$x+1=2\sqrt{x+1}$$
$$(x+1)^2=(2\sqrt{x+1})^2$$
$$x^2+2x+1=4(x+1)$$
$$x^2+2x+1=4x+4$$
$$x^2-2x-3=0$$
$$(x-3)(x+1)=0$$

$\Rightarrow x=-1$ or $x=3$

Check:
$x=-1$: $\sqrt{2x+3}-\sqrt{x+1}=\sqrt{2(-1)+3}-\sqrt{(-1)+1}=\sqrt{1}-\sqrt{0}=1$
$x=3$: $\sqrt{2x+3}-\sqrt{x+1}=\sqrt{2(3)+3}-\sqrt{(3)+1}=\sqrt{9}-\sqrt{4}=1$. Then the original equation has the solution set $\{-1,\ 3\}$.

Example 5: Find the real solution: $(2-x)^2+(2-x)-20=0$

Solution: Let $u=2-x$, then the original equation becomes

$$u^2+u-20=0.$$

Factor

$$(u-4)(u+5)=0$$
$$u=4 \text{ or } u=-5.$$

But remember that we want to solve for x. Because $u=2-x$, then we have

$$2-x=4 \text{ or } 2-x=-5$$
$$x=-2 \text{ or } x=7.$$

Thus, the solution set is $\{-2,\ 7\}$. Check this solution!

Example 6: Solve by factoring: $x^3+6x^2-7x=0$

Solution: Factor out the common factor x.

$$x(x^2+6x-7)=0$$
$$x(x+7)(x-1)=0$$
$$x=-7,\ x=0,\ x=1$$

Check the solution!

Example 7: Solve by factoring: $x^3+4x^2-x-4=0$

Solution: Do you recall the method of factoring by grouping?

$$(x^3+4x^2)-(x+4)=0$$
$$x^2(x+4)-(x+4)=0$$
$$(x+4)(x^2-1)=0$$
$$(x+4)(x-1)(x+1)=0$$
$$x=-1,\ x=-4,\ x=1.$$

Check the solution!

Group Exercises 3

Radical Equations and Factoring Methods

(1) Find the real solutions of each equation:

(a) $\sqrt{2t-1}=1$

(b) $\sqrt{3t+4}=-6$

(c) $\sqrt{15-2x}=x$

(d) $\sqrt[3]{1-2x}-3=0$

(e) $3x-\sqrt{1+3x}=1$

(f) $\sqrt{3x+1}-\sqrt{x-1}=2$

(2) Indicate an appropriate substitution of variable and solve each equation:

(a) $x^4 - 10x^2 + 25 = 0$

(b) $3(1 - y)^2 + 5(1 - y) + 2 = 0$

(3) Find the real solution of each equation by factoring:

(a) $4x^3 - 3x^2 = 0$

(b) $4x^3 + 2x^2 + 3x = 0$

(c) $x^3 - 3x^2 - x + 3 = 0$

Practice Exercises 3

Radical Equations and Factoring Methods

(1) Find the real solutions of each equation:

(a) $\sqrt{3t+4} = 2$

(b) $\sqrt{5t+3} = -2$

(c) $\sqrt{12-x} = x$

(d) $\sqrt[3]{1-2x} - 1 = 0$

(e) $x - \sqrt{13-x} = 3$

(f) $\sqrt{4-4x} - \sqrt{x+4} = 3$

(2) Indicate an appropriate substitution of variable and solve each equation:

(a) $x^4 - 5x^2 + 4 = 0$

(b) $2(s+1)^2 - 5(s+1) = 3$

(3) Find the real solution of each equation by factoring:

(a) $x^4 - x^2 = 0$

(b) $x^3 + x^2 - x - 1 = 0$

(c) $x^3 + x^2 - 20x = 0$

(d) $4x^3 = 3x^2$

Chapter 4

Inequalities and Interval Notation

Intervals: Let a and b be two real numbers with $a < b$.

A **closed interval**, denoted by $[\mathbf{a},\ \mathbf{b}]$, consists of all real numbers x for which $a \leq x \leq b$.

An **open interval**, denoted by $(\mathbf{a},\ \mathbf{b})$, consists of all real numbers x for which $a < x < b$.

The **half-open interval**, or **half-closed interval**, denoted by $[\mathbf{a},\ \mathbf{b})$, consists of all real numbers x for which $a \leq x < b$.

The **half-open interval**, or **half-closed interval**, denoted by $(\mathbf{a},\ \mathbf{b}]$, consists of all real numbers x for which $a < x \leq b$.

The **interval** $[\mathbf{a},\ \infty)$, consists of all real numbers x for which $x \geq a$ $(a \leq x < \infty)$.

The **interval** $(\mathbf{a},\ \infty)$, consists of all real numbers x for which $x > a$ $(a < x < \infty)$.

The **interval** $(-\infty,\ \mathbf{a}]$, consists of all real numbers x for which $x \leq a$ $(-\infty < x \leq a)$.

The **interval** $(-\infty,\ \mathbf{a})$, consists of all real numbers x for which $x < a$ $(-\infty < x < a)$.

The **interval** $(-\infty,\ \infty)$, consists of all real numbers x $(-\infty < x < \infty)$.

Example 1: Write each inequality using interval notation.

(1) $-3 \leq x \leq 4$

Solution: $-3 \leq x \leq 4$ describes all numbers x between -3 and 4, inclusive. In interval notation, we write $[-3,\ 4]$.

(2) $0 < x < 12$

Solution: In interval notation, $0 < x < 12$ is written as $(0,\ 12)$.

(3) $x < -5$

Solution: In interval notation, $x < -5$ is written as $(-\infty,\ -5)$.

(4) $x \geq 0$

Solution: In interval notation, $x \geq 0$ is written as $(0,\ \infty)$.

(5) $-3 < x \leq -1$

Solution: In interval notation, $-3 < x \le -1$ is written as $(-3,\ -1]$.

Example 2: Write each interval as an inequality involving x.

(1) $[-1,\ 3)$

Solution: $[-1,\ 3)$ consists of all numbers x for which $-1 \le x < 3$.

(2) $[-1,\ 1]$

Solution: $[-1,\ 1]$ consists of all numbers x for which $-1 \le x \le 1$.

(3) $(-\infty,\ 5]$

Solution: $(-\infty,\ 5]$ consists of all numbers x for which $-\infty < x \le 5$.

(4) $(-7,\ \infty)$

Solution: $(-7,\ \infty)$ consists of all numbers x for which $-7 < x < \infty$.

(5) $(13,\ 20)$

Solution: $(13,\ 20)$ consists of all numbers x for which $13 < x < 20$.

Properties of Inequalities

If $a < b$, then $a + c < b + c$.
If $a > b$, then $a + c > b + c$.
If $a < b$ and if $c > 0$, then $ac < bc$.
If $a < b$ and if $c < 0$, then $ac > bc$.
If $a > b$ and if $c > 0$, then $ac > bc$.
If $a > b$ and if $c < 0$, then $ac < bc$.
If $a > 0$, the $\frac{1}{a} > 0$.
If $a < 0$, then $\frac{1}{a} < 0$.

Example 3: Solve each inequality and graph the solution set.

(1) $3x - 5 > 10$

Solution: $x > 5$

(2) $2 - 5x \geq -8$

Solution: $x \leq 2$

(3) $4(x-3) \leq 3(x-2)$

Solution: $x \leq 6$

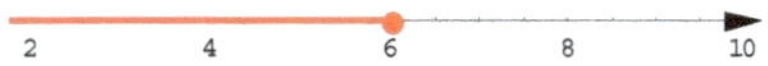

(4) $5 < 3x - 1 \leq 11$

Solution: $2 < x \leq 4$

(5) $\dfrac{1}{5} < \dfrac{x-2}{3} < \dfrac{3}{5}$

Solution: $\dfrac{13}{5} < x < \dfrac{19}{5}$

(6) $\dfrac{3}{2}(x+1) + \dfrac{5}{6} \geq \dfrac{1}{2}(2x-1) + 4$

Solution: $x \geq \dfrac{7}{3}$

(7) $(2x-4)^{-1} > 0$

Solution: $x > 2$

Group Exercises 4

Inequalities and Interval Notation

(1) Write each inequality using interval notation.

(a) $0 \leq x \leq 4$

(b) $-4 < x \leq 1$

(c) $-5 \leq x < 1$

(d) $3 < x < 7$

(e) $x < 13$

(f) $x > 1$

(g) $x \leq -2$

(h) $x \geq 0$

(2) Write each interval as an inequality involving x.

(a) $[2,\ 5]$

(b) $(-2,\ -1)$

(c) $(0,\ 5]$

(d) $[-3,\ 0)$

(e) $[-7, \infty)$

(f) $(2, \infty)$

(g) $(-\infty, 1]$

(h) $(-\infty, -3)$

(3) Solve each of the following Inequalities:

(a) $2 - 3x \leq 5$

(b) $4 - 3(1 - x) \leq 3$

(c) $2x - 2 \geq 3 + x$

(d) $\frac{1}{2}(x - 4) > x + 8$

(e) $-3 < \dfrac{2x-1}{4} < 0$

(f) $\dfrac{1}{2} \le \dfrac{x+1}{3} < \dfrac{3}{4}$

(g) $(4x+2)^{-1} < 0$

Problem I. In your Math 120 class, you have scores of 68, 82, 87, and 89 on the first four of five tests. To get a grade of **B**, the average of the first five test scores must be greater than or equal to 80 and less than 90. Solve an inequality to find the range of the score that you need on the last test to get a **B**.

Problem II. Commonwealth Edison Company's charge for electricity in May 2003 is 8.275 cents per kilowatt-hour. In addition, each monthly bill contains a customer charge of \$7.58. If last year's bills ranged from a low of \$63.47 to a high of \$214.53, over what range did usage vary (in kilowatt-hours)? .

Practice Exercises 4

Inequalities and Interval Notation

(1) Write each inequality using interval notation.

(a) $-10 \leq x \leq -2$

(b) $1 < x \leq 7$

(c) $0 \leq x < 3$

(d) $-2 < x < 2$

(e) $x < 5$

(f) $x > -1$

(g) $x \leq 0$

(h) $x \geq -4$

(2) Write each interval as an inequality involving x.

(a) $[-2,\ 0]$

(b) $(-2,\ 3)$

(c) $(-4,\ 4]$

(d) $[-1,\ 0)$

(e) $[7, \infty)$

(f) $(-2, \infty)$

(g) $(-\infty, -1]$

(h) $(-\infty, 3)$

(3) Solve each of the following Inequalities and graph the solution set:

(a) $-5x + 2 > -8$

(b) $-3(1 - x) < 12$

(c) $8 - 4(2 - x) \leq -2x$

(d) $1 - 2x \leq 3$

(e) $3(2a - 1) > 4(2a - 3)$

(f) $\frac{1}{4}(3x+2) - 1 \leq -\frac{1}{2}(x-3) + \frac{3}{4}$

(g) $0 < \frac{3x+2}{2} < 4$

(h) $\frac{1}{3} < \frac{x+1}{2} \leq \frac{2}{3}$

(i) $(2x-1)^{-1} > 0$

Problem I. An appliance salesperson is paid \$30 per day plus \$25 for each appliance sold. How many appliances must be sold for the salesperson's income to exceed \$130 per day?

Problem II. A student has grades of 42 and 70 on the first two tests in Math 120 of the semester. If an average of 70 is required to obtain a **C** grade, what is the minimum score the student must achieve on the third exam to obtain a **C**?

Problem III. A real estate agent agrees to sell a house according to the following commission schedule: \$2000 plus 25% of the selling price in excess of \$180, 000. Assuming that the house will sell at some price between \$180, 000 and \$260, 000 inclusive, over what range does the agent's commission vary?

Chapter 5

Equations and Inequalities Involving Absolute Value

Suppose we are interested in the distance between the origin and the points 2 and -2 on the real number line. Each of these points is 2 units from the origin, so the distance between 2 and -2 is 4 units. That is, the distance is independent of the direction and is nonnegative.

In general, when we are interested in the magnitude of a number x, we do not care about the direction or the sign of x. Therefore, the concept of the the **absolute value** is introduced, and we write $|x|$.

$$|x| = \begin{cases} x & \text{if } x \geq 0 \\ -x & \text{if } x < 0 \end{cases}$$

Example 1:

(1) $|\ 2\ | = 2$

(2) $|-2| = 2$

(3) $|\ 0\ | = 0$

(4) $|\ 4-1\ | = 3$

(5) $|\ 3-8\ | = |-5| = 5$

If a is a positive number and u is any algebraic expression, then

$$\boxed{|\ u\ | = a \text{ is equivalent to } u = a \text{ or } u = -a}$$

Example 2: Solving an equation or an inequality involving absolute value.

(1) Solve the equation $|2x+3| = 5$.

Solution: By the definition above, we have two cases:

Case 1: $2x+3 \geq 5$

$|2x+3| = 2x+3 = 5$

$2x = 2$

$x = 1$

Case 2: $2x+3 < 5$

$|2x+3| = -(2x+3) = 5$

$-2x-3 = 5$

$x = -4$

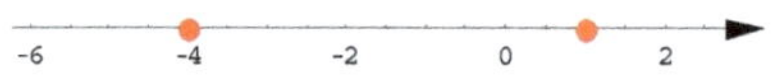

(2) Solve the equation $3|-4x-3|+4 = 31$.

Solution: First, subtract 4 from both sides.
$3|-4x-3| = 27$
then divide both sides by 3
$|-4x-3| = 9.$
By the definition above, we have two cases:

Case 1:
$-4x - 3 = 9$
$-4x = 12$
$x = -3$

Case 2:
$-4x - 3 = -9$
$-4x = -6$
$x = \frac{3}{2}$

(3) Solve the inequality $|x-2| + 2 < 3$

Solution:
$|x-2| < 1$
$-1 < x - 2 < 1$
$1 < x < 3.$

The solution set is $\{x|\ 1 < x < 3\}$, that is, all real numbers x in the open interval $(1,\ 3)$. See the graph below.

If a is a positive number and u is any algebraic expression, then

If $\mid u \mid < a$ is equivalent to $-a < u < a$
If $\mid u \mid \leq a$ is equivalent to $-a \leq u \leq a$
If $\mid u \mid > a$ is equivalent to $u > a$ or $u < -a$
If $\mid u \mid \geq a$ is equivalent to $u \geq a$ or $u \leq -a$

(4) Solve and graph the solution set: $|x-3| \geq 2$.

Solution:
$x - 3 \geq 2$ or $x - 3 \leq -2$
$x \geq 5$ or $x \leq 1$.

The solution set consists of all numbers x for which $x \geq 5$ or $x \leq 1$, that is, all x in the intervals $(-\infty,\ 1]$ or $[5,\ \infty)$.

(5) Solve and graph the solution set: $|1-4x|-7<-2$.

Solution:

$$|1-4x|<5$$
$$-5<1-4x<5$$
$$-6<-4x<4$$
$$\frac{-6}{-4}>x>\frac{4}{-4}$$
$$\frac{3}{2}>x>-1 \text{ or } -1<x<\frac{3}{2}.$$

The solution set consists of all numbers x for which $-1<x<\frac{3}{2}$, that is, all x in the interval $(-1,\ \frac{3}{2})$.

Group Exercises 5

Solving Absolute Value Equations and Inequalities

(1) Solve each equation.

(a) $|2x - 1| = 2$

(b) $|1 - 4t| + 8 = 13$

(c) $|u - 2| = -\dfrac{1}{2}$

(d) $|x^2 - 16| = 0$

(e) $|\dfrac{x}{2} - \dfrac{1}{3}| = 1$

(2) Solve each inequality. Express your answer using set notation or interval notation.

(a) $|2t - 2| \leq 4$

(b) $|x + 4| + 3 < 5$

(c) $|1 - 2x| - 4 < -1$

(d) $|x + 4| \geq 2$

(e) $|2 - 3x| > 1$

(f) $-|1-2x| \geq -3$

Problem I. "Normal" human body temperature is 98.6°F. If a temperature x that differs from normal by at least 1.5° is considered unhealthy, write the condition for unhealthy temperature x as an inequality involving an absolute value, and solve for x.

Practice Exercises 5

Solving Absolute Value Equations and Inequalities

(1) Solve each equation.

(a) $|2x - 4| = 2$

(b) $|2t + 2| = 0$

(c) $3|-4x - 3| = 27$

(d) $|2x + 1| - 3 = -2$

(e) $3 - |2x + 4| = 1$

(2) Solve each inequality. Express your answer using set notation or interval notation.

(a) $|\frac{x}{3} + \frac{2}{5}| < 2$

(b) $|3t - 2| \leq 4$

(c) $\frac{|2x - 1|}{4} < 2$

(d) $|x - 3| \geq 4$

(e) $|3x| > 0$

(f) $|3x + 2| + 1 > 0$

Problem I. In the United States, normal house-hold voltage is 115 volts. However, it is not uncommon for actual voltage to differ from normal voltage by at most 5 volts. Express this situation as an inequality involving an absolute value. Use x as the actual voltage and solve for x.

Chapter 6

Lines and Their Graphs

Theorem. (The Distance Formula.) The distance between two points $P_1(x_1,\ y_1)$ and $P_2(x_2,\ y_2)$, denoted by $d(P_1,\ P_2)$, is

$$\boxed{d(P_1, P_2) = \sqrt{(x_2 - x_1)^2 + (y_2 - y_1)^2}}$$

Example 1: Find the distance between the points $(2,\ 2)$ and $(-2,\ -1)$.

Solution: $d^2 = 16 + 9 = 25$, so $d = \sqrt{25} = 5$.

Example 2: Find the distance between the points $P(-1,\ 3)$ and $Q(4,\ -2)$.

Solution: $d = \sqrt{(4-(-1))^2 + (-2-3)^2} = \sqrt{5^2 + (-5)^2} = \sqrt{25+25} = \sqrt{50} = 5\sqrt{2}$

Example 3: Show that the triangle with vertices $A(-2,\ 3)$, $B(3,\ -2)$ and $C(6,\ 1)$ is a right triangle.

Solution: Compute the lengths of the three sides of the triangle.

$$\begin{aligned} \overline{AB} &= \sqrt{(3+2)^2 + (-2-3)^2} = \sqrt{50} \\ \overline{BC} &= \sqrt{(6-3)^2 + (1+2)^2} = \sqrt{18} \\ \overline{AC} &= \sqrt{(6+2)^2 + (1-3)^2} = \sqrt{68} \end{aligned}$$

If the Pythagorean Theorem holds, then triangle ABC is a right triangle.

$$(\overline{AC})^2 = (\overline{AB})^2 + (\overline{BC})^2 \qquad \text{since} \quad 68 = 50 + 18$$

Theorem. (The Midpoint Formula.) The midpoint $M = (x,\ y)$ of the line segment from $P_1(x_1,\ x_2)$ to $P_2(x_2,\ y_2)$ is

$$\boxed{M = (x,\ y) = (\frac{x_1 + x_2}{2},\ \frac{y_1 + y_2}{2})}$$

Example 4: Find the coordinates of the midpoint of the line segment with endpoints $P(-2,\ -3)$ and $Q(1,\ 2)$

Solution: The coordinates $(x,\ y)$ of the midpoint M are

$$\begin{aligned} x &= \frac{x_1 + x_2}{2} = \frac{-2+1}{2} = \frac{-1}{2} \\ y &= \frac{y_1 + y_2}{2} = \frac{-3+2}{2} = \frac{-1}{2} \end{aligned}$$

Slope of a Line

Consider the staircase, each step contains exactly the same horizontal distance (**run**) and the same vertical distance (**rise**). The ratio of the rise to the run , called the *slope*. The slope of a straight line is also defined to be the **rate of change** of the y variable with respect to the x variable. The slope measures the steepness of a straight line and it is always a constant (why?).

Let $P(x_1,\ y_1)$ and $Q(x_2,\ y_2)$ be two distinct points. If $x_1 \neq x_2$, the **slope** m of the nonvertical line L containing the points P and Q is defined by the formula

$$\boxed{m = \frac{y_2 - y_1}{x_2 - x_1} = \frac{\Delta y}{\Delta x} = \frac{\text{Rise}}{\text{Run}}, \quad x_1 \neq x_2}$$

where Δ means " the change in". See the figure below.

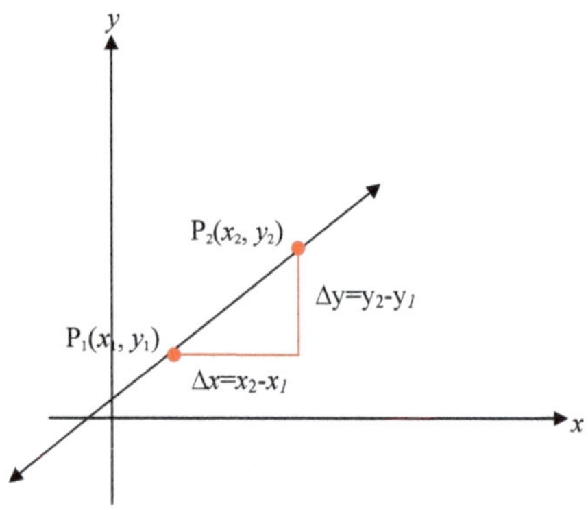

Example 5: If $x_1 = x_2$, L is a vertical line and the slope m is undefined. The equation of the line in the graph below is $x = 3$.

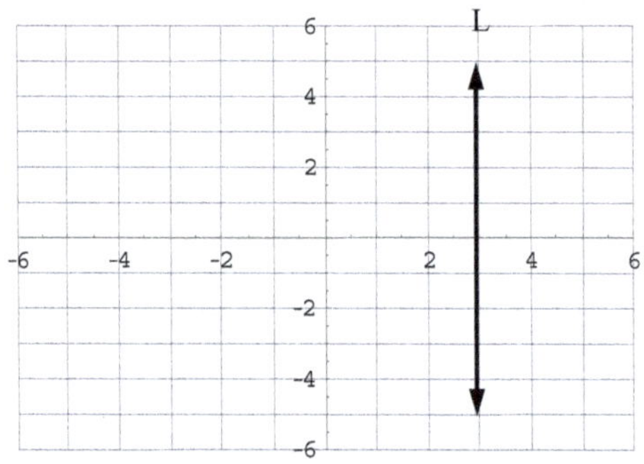

Example 6: If $y_1 = y_2$, L is a horizontal line and the slope $m = 0$. The equation of the line in the graph below is $y = 4$.

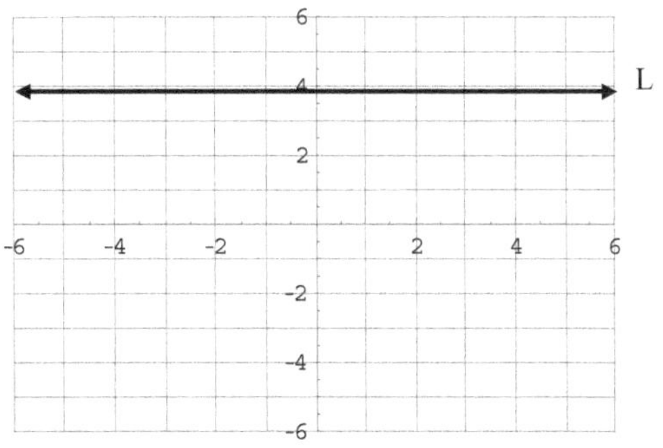

Example 7: Determine the slope of the line containing the points $(1,\ 3)$ and $(-1,\ -7)$.

Solution: Using the slope formula, $m = \dfrac{-7-3}{-1-1} = \dfrac{-10}{-2} = 5$

Example 8: If the slope of a line L is $m = \dfrac{-1}{2}$ and the point $P(2,\ 5)$ is on the line, use this information to find another two points on the line.

Solution: Plot the point $(2,\ 5)$ and use the slope $\dfrac{-1}{2}$ to locate another two points on the graph on the line. If we move 2 units to the left and 1 unit up, we will get to the point $(0,\ 6)$. From this point, move again 2 units to the left and 1 unit up to get to the point $(-2,\ 7)$.

Example 9: Graph the line containing the point $P(-3,\ 4)$ and having slope m 2.

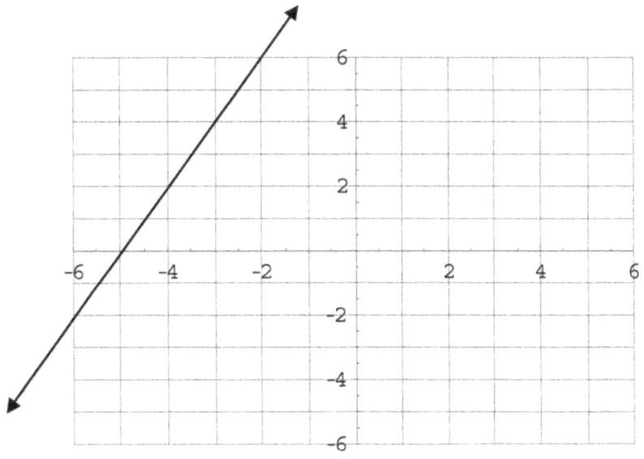

Group Exercises 6

Lines and Their Graphs

(1) Plot each point in the xy−plane.

(a) $A(-3,\ 2)$
(b) $B(0,\ 3)$
(c) $C(5,\ 0)$
(d) $D(-3,\ -2)$
(e) $E(-1,\ 0)$

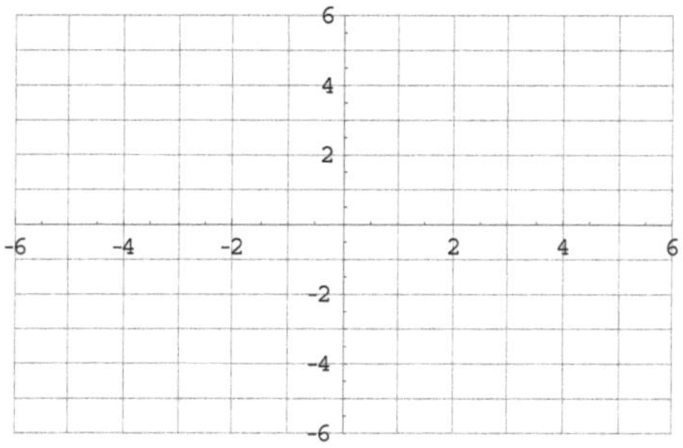

(2) Find the distance $d(P_1, P_2)$ between the given points P_1 and P_2.

(a) $P_1(3,\ -4)$; $P_2(5,\ 4)$

(b) $P_1(-4,\ 5)$; $P_2(-2,\ 3)$

(c) $P_1(-4,\ -5)$; $P_2(-5,\ -1)$

(d) $P_1(-1,\ 0)$; $P_2(2,\ 4)$

(3) Plot each point and form the triangle ABC. Verify that the triangle is a right triangle. Find its area.

(a) $A(-2,\ 5)$; $B(1,\ 3)$; $C(-1,\ 0)$

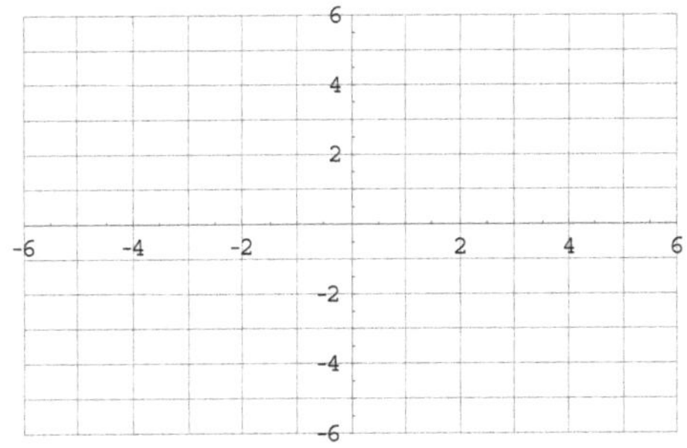

(b) $A(4,\ -3)$; $B(0,\ -3)$; $C(4,\ 2)$

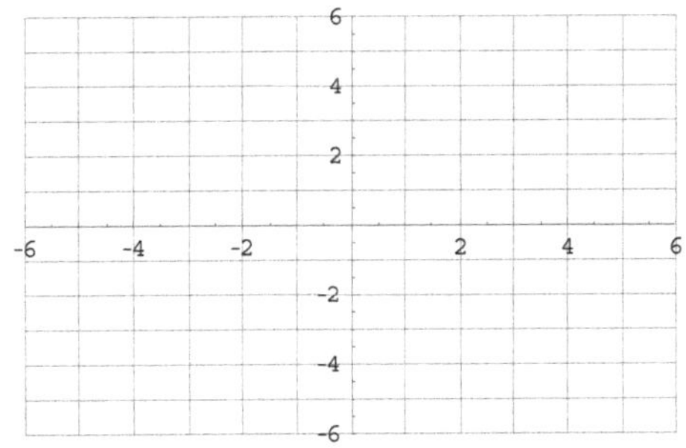

(4) Find the midpoint of the line segment joining the points P_1 and P_2.

(a) $P_1(5,\ -4)$; $P_2(3,\ 2)$

(b) $P_1(4,\ -3)$; $P_2(6,\ 1)$

(c) $P_1(-3,\ 0)$; $P_2(2,\ -4)$

(d) $P_1(5,\ 4)$; $P_2(2,\ 1)$

(5) Plot each pair of points and determine the slope of the line.

(a) $(2,\ 3)$, $(4,\ 0)$

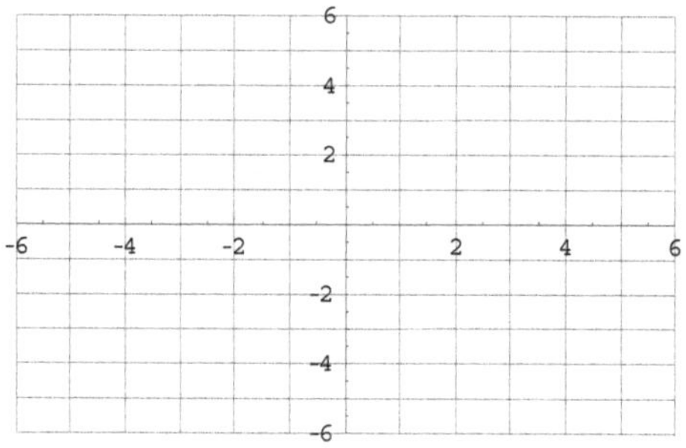

(b) $(2,\ 0)$, $(2,\ 2)$

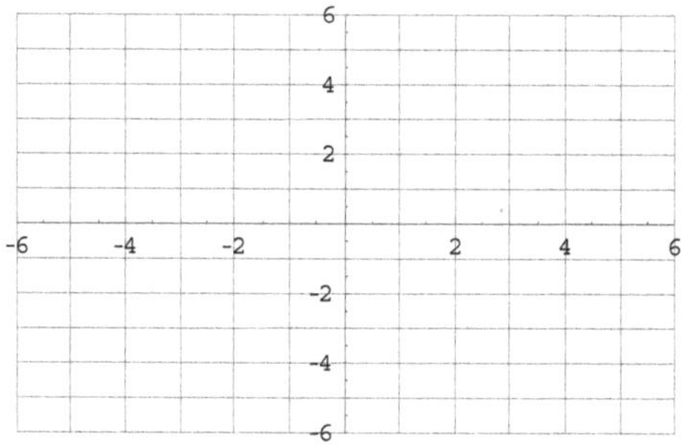

(c) $(-4,\ 1)$, $(-1,\ -2)$

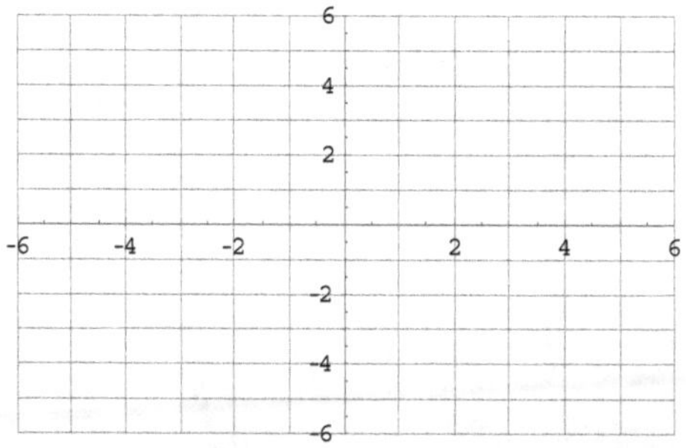

(6) Graph the line containing the point P and having slope m.

(a) $P(1,\ 2)$; $m = 3$

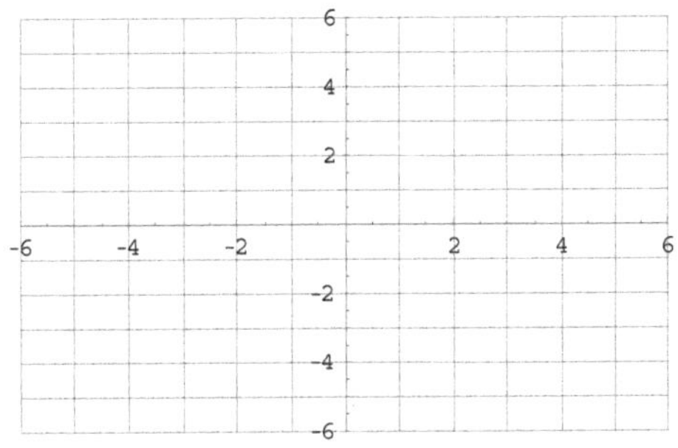

(b) $P(0,\ 3)$; $m =$ undefined

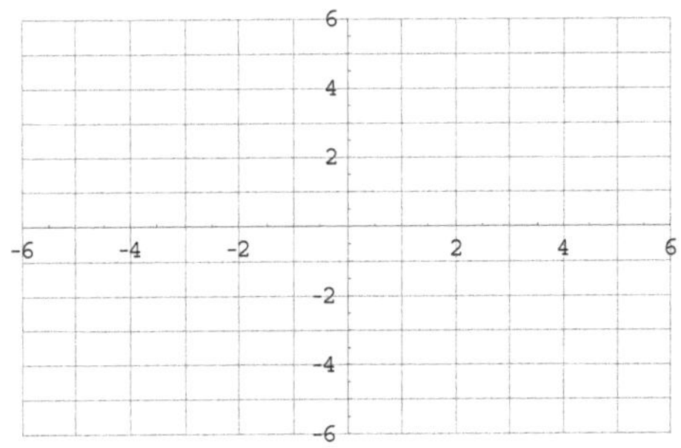

(7) The slope and a point on a line are given. Use this information to locate two additional points on the line.

(a) $m = 2$; point $(-2,\ 3)$

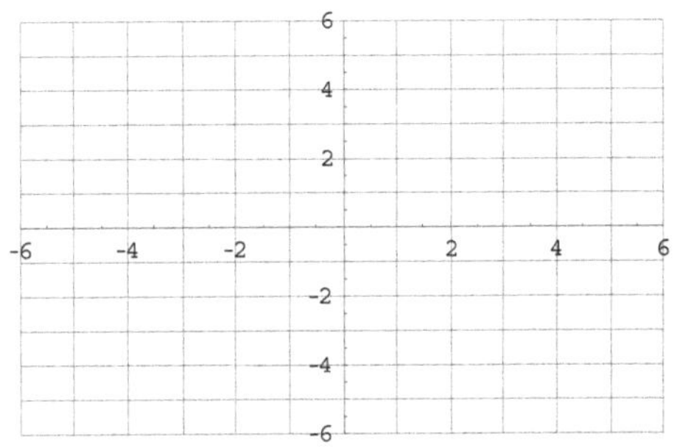

(b) $m = -2$; point $(-2,\ -3)$

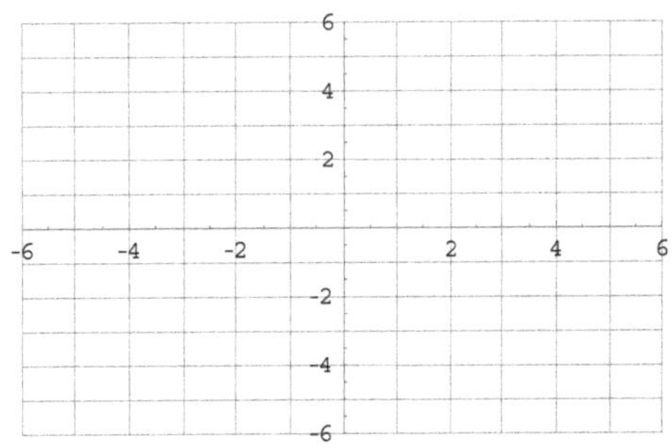

Practice Exercises 6

Lines and Their Graphs

(1) Plot each point in the $xy-$plane.

(a) $A(-1,\ 0)$

(b) $B(0,\ -1)$

(c) $C(-3,\ -3)$

(d) $D(2,\ 2)$

(e) $E(\frac{1}{2},\ \frac{3}{2})$

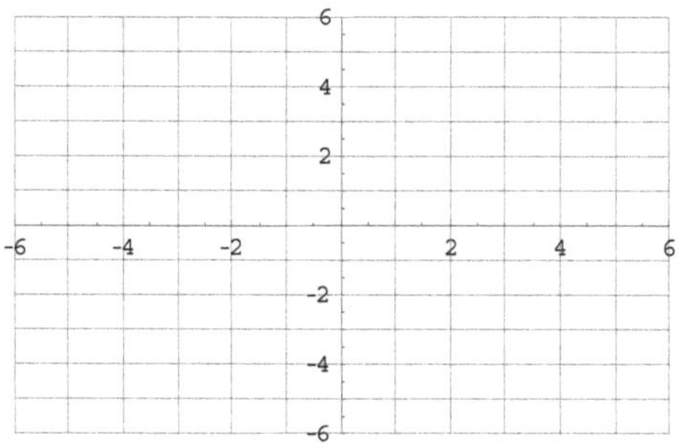

(2) Find the distance $d(P_1, P_2)$ between the given points P_1 and P_2.

(a) $P_1(4,\ -3)$; $P_2(6,\ 4)$

(b) $P_1(-2,\ -3)$; $P_2(4,\ 2)$

(c) $P_1(-1,\ 1)$; $P_2(2,\ 2)$

(d) $P_1(-3,\ 0)$; $P_2(2,\ -4)$

(3) Plot each point and form the triangle ABC. Verify that the triangle is a right triangle. Find its area.

(a) $A(2,\ -3)$; $B(-1,\ -1)$; $C(3,\ 4)$

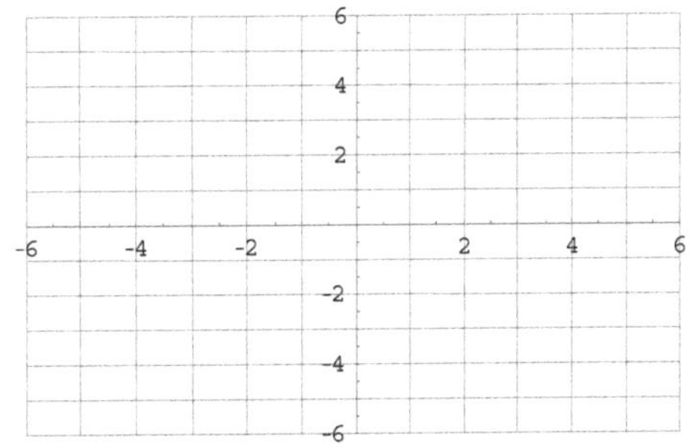

(b) $A(1,\ -2)$; $B(5,\ 2)$; $C(2,\ 1)$

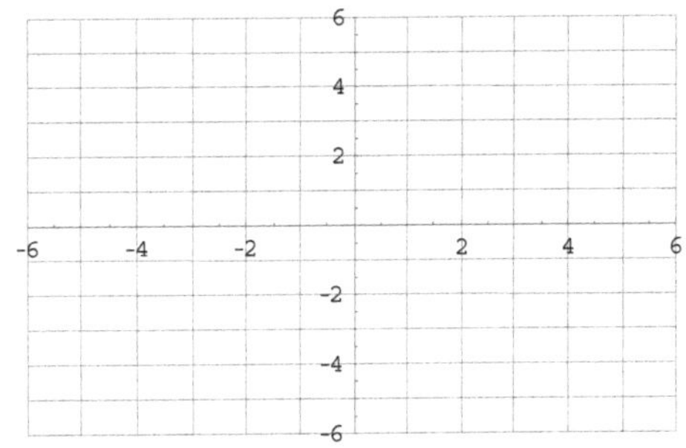

(4) Find the mid-point of the line segment joining the points P_1 and P_2.

(a) $P_1(2,\ -3)$; $P_2(4,\ 2)$

(b) $P_1(-4,\ -3)$; $P_2(2,\ 2)$

(c) $P_1(-3,\ 2)$; $P_2(6,\ 0)$

(d) $P_1(a,\ b)$; $P_2(0,\ 0)$

(5) Plot each pair of points and determine the slope of the line.

(a) $(-2,\ 3)$, $(2,\ 1)$

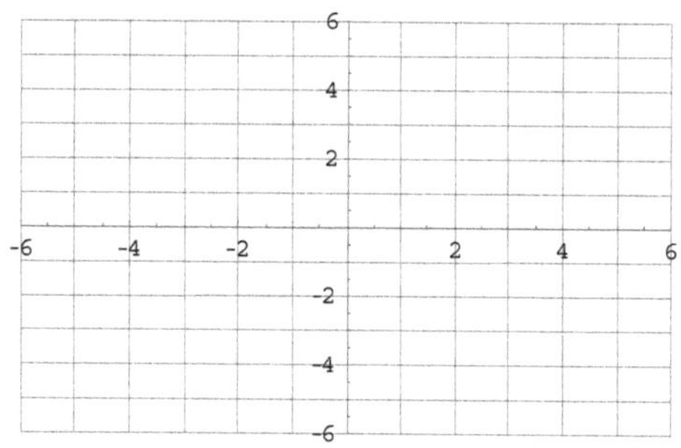

(b) $(-2,\ 3)$, $(0,\ 0)$

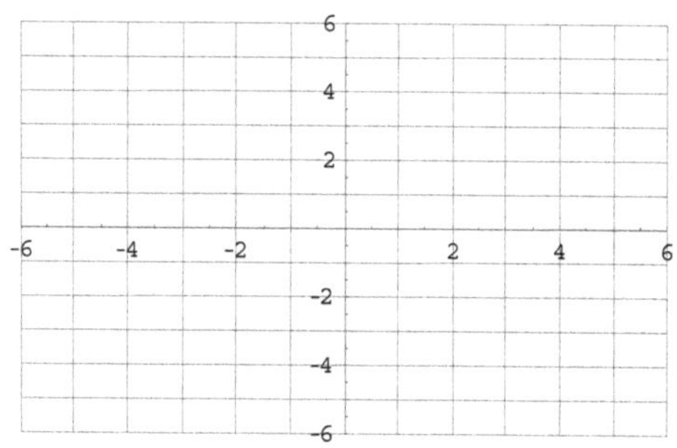

(c) $(2,\ 4)$, $(-3,\ 4)$

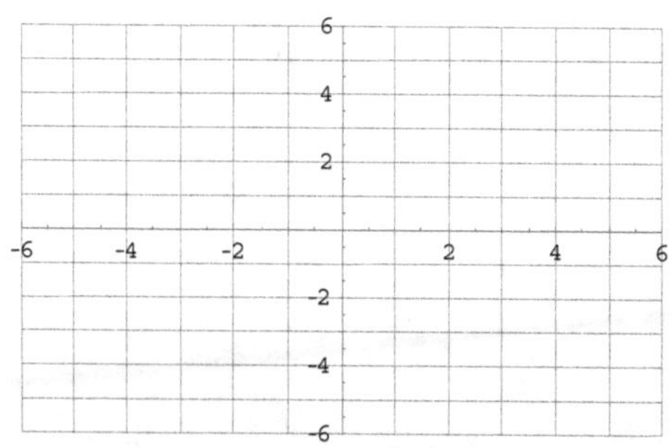

(6) Graph the line containing the point P and having slope m.

(a) $P(2,\ 4);\ m = \dfrac{-3}{4}$

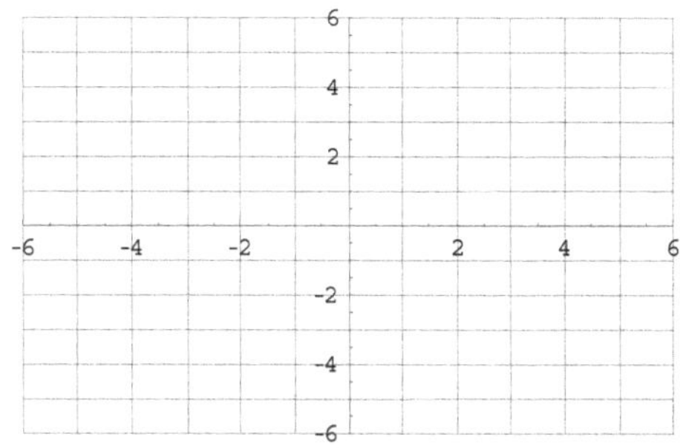

(b) $P(2,\ -4);\ m = 0$

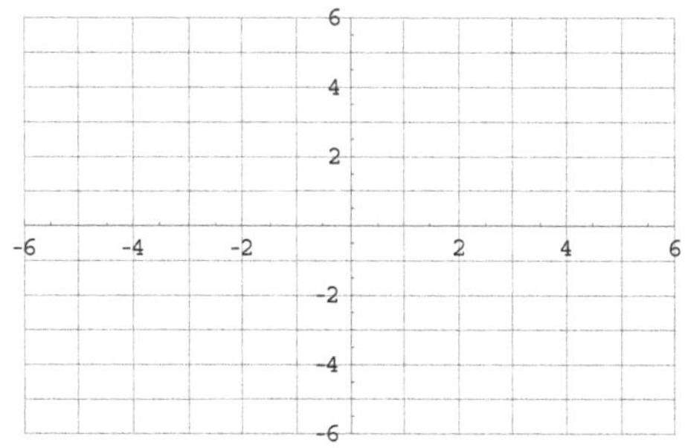

(7) The slope and a point on a line are given. Use this information to locate two additional points on the line.

(a) $m = -1$; point $(4,\ 1)$

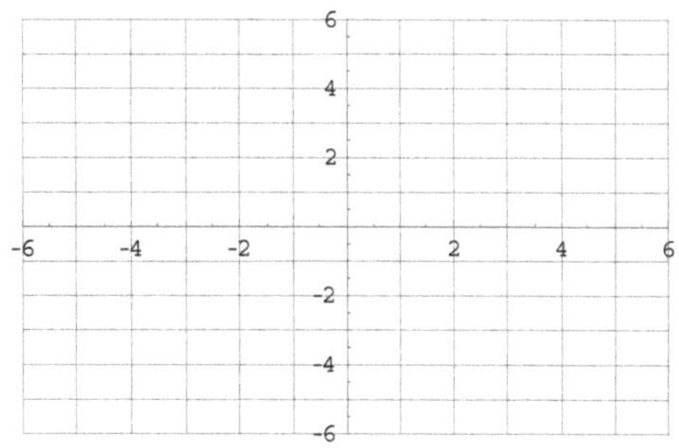

(b) $m = \dfrac{4}{3}$; point $(-3,\ 2)$

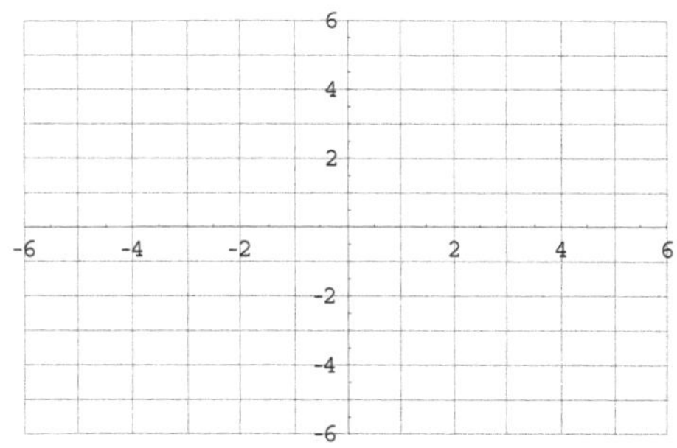

Chapter 7

Equations of Lines and Applications

(1) Equation of a Vertical Line.

The equation of a vertical line through the point $(a,\ b)$ is $x = a$. The slope of a vertical line is undefined.

(2) Equation of a Horizonal Line.

The equation of a horizontal line through the point $(a,\ b)$ is $y = b$. The slope of a horizontal line is 0.

Example 1: Find the equation of the vertical and horizontal line through the point $(-3,\ 2)$.

Solution: The vertical line has the equation $x = -3$. The horizontal line has the equation $y = 2$.

(3) Point-Slope Form of an Equation of a Line.

Let (x_1, y_1) be a point on a line L with slope m, and for any other point $(x,\ y)$ on the line L, by the slope formula, we have $m = \dfrac{y - y_1}{x - x_1}$ or $y - y_1 = m(x - x_1)$.

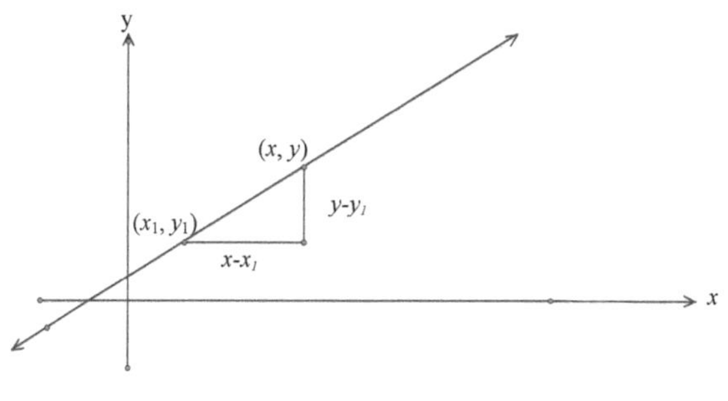

An equation of a nonvertical line with slope m that passes through the point $(x_1,\ y_1)$ is $y - y_1 = m(x - x_1)$.

Example 2: Find an equation of the line that passes through the points $(1,\ -2)$ and $(-3,\ 4)$.

Solution: First find the slope. Let $x_1 = 1$, $y_1 = -2$, $x_2 = -3$ and $y_2 = 4$. Then

$$m = \frac{y_2 - y_1}{x_2 - x_1} = \frac{4-(-2)}{-3-1} = \frac{6}{-4} = -\frac{3}{2}$$

Now use the point-slope form with $m = -\frac{3}{2}$ and $(x_1,\ y_1) = (1,\ -2)$.

$$\begin{aligned} y - y_1 &= m(x - x_1) \\ y - (-2) &= -\frac{3}{2}(x-1) \\ y + 2 &= -\frac{3}{2}x + \frac{3}{2} \\ y &= -\frac{3}{2}x - \frac{1}{2} \end{aligned}$$

(4) Slope-Intercept Form of an Equation of a Line

An equation of a line L with slope m and $y-$intercept b is $y = mx + b$.

Example 3: Find the slope m and $y-$intercept b of the line with equation $2x-3y = 6$.

Solution: First, write the equation in the form $y = mx + b$ by solving the equation for y.

$$y = \frac{2}{3}x - 2.$$

The coefficient of x, $-\frac{2}{3}$, is the slope m, and the $y-$intercept b, is -2.

(5) The General Form of an Equation of a Line

The equation $Ax + By = C$ is said to be in **general form**, where A, B, and C are real numbers and A and B are not both 0.

Exercise: What is the equation when $A = 0$?

Exercise: What is the equation when $B = 0$?

Group Exercises 7

Equations of Lines and Applications

(1) Find an equation of each line. Express your answer using the slope-intercept form of the equation of a line.

(a)

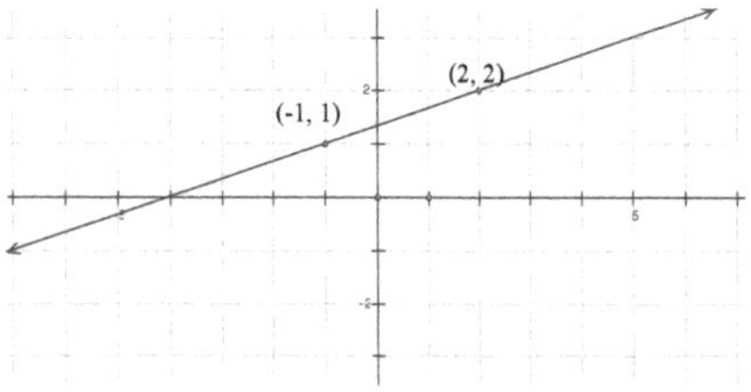

(b)

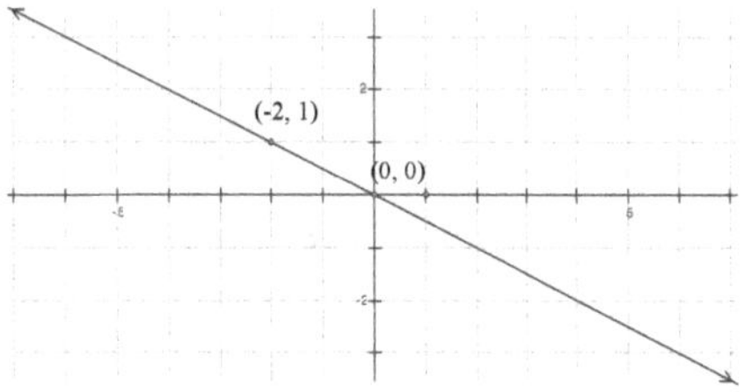

(2) Find an equation of the line with the given properties. Express your answer using the general form of the equation of a line.

(a) Slope $m = 2$; containing the point $(-1,\ 3)$

(b) Containing the points $(1,\ 3)$ and $(1,\ -7)$

(c) Slope $m = -3$; $y-$intercept $b = 0$

(d) $x-$intercept $= 4$; $y-$intercept $= -4$

(e) Slope $m = 0$; $y-$intercept $b = 2$

(3) Find the slope and $y-$intercept of each line.

(a) $y = 3x - 1$

(b) $2x - 5y = -3$

(c) $3x + 4y = -2$

(d) $y - 4 = 0$

(e) $x + 5 = 0$

(4) The equation of a line L is given. Find the slope of a line that is (a) parallel to L and (b) perpendicular to L.

(a) $y = 7$

(b) $x = 6$

(c) $3x + 5y - 20 = 0$

(d) $y = -3x$

(5) Find an equation of the line that is parallel to the line $2x - y = -2$ and containing the point $(0,\ 0)$.

Problem I. The Celsius (C) and Fahrenheit (F) temperature scales are related by a linear equation. Water boils at 212°F or 100°C, and freezes at 32°F or 0°C.

(1) Write a linear equation expressing F in terms of C.

(2) Use the equation to find the Celsius temperature when the temperature is 70°F.

Problem II. The college bookstore sells a textbook that costs \$40 for \$45, and a textbook that costs \$42 for \$47. If the markup policy of the bookstore is linear, write a linear equation that relates sales price S and cost C. Then use the equation to find the cost of a book that sells for \$52.

Practice Exercises 7

Equations of Lines and Applications

(1) Find an equation of each line. Express your answer using the slope-intercept form of the equation of a line.

(a)

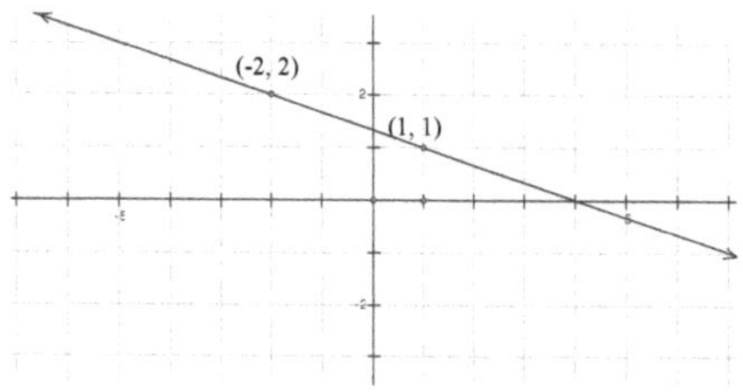

(b)

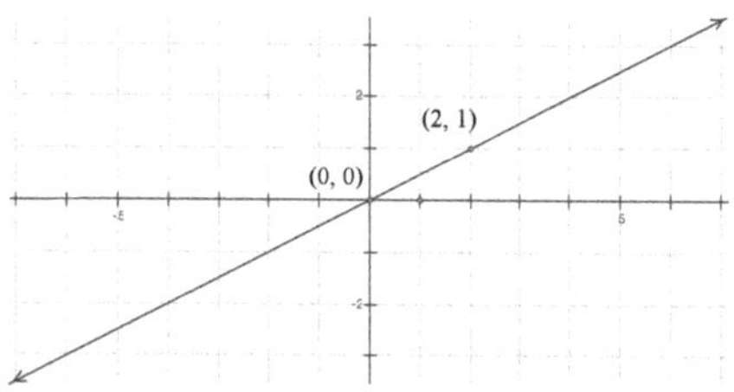

(2) Find an equation of the line with the given properties. Express your answer using the general form of the equation of a line.

(a) Slope $m = -\frac{1}{2}$; containing the point $(4,\ -3)$

(b) Containing the points $(-3,\ 4)$ and $(2,\ 5)$

(c) Slope $m = -2$; $y-$intercept $b = 0$

(d) $x-$intercept $= -4$; $y-$intercept $= 4$

(e) Slope $m = 0$; $y-$intercept $b = -\frac{1}{2}$

(3) Find the slope and $y-$intercept of each line.

(a) $y = -3x + 4$

(b) $x + 2y = 4$

(c) $3x + 2y - 6 = 0$

(d) $x - 3 = 0$

(e) $y + 3 = 0$

(4) The equation of a line L is given. Find the slope of a line that is (a) parallel to L and (b) perpendicular to L.

(a) $y = -5$

(b) $x = 2$

(c) $2x - 4y + 5 = 0$

(d) $y = 6x$

(5) Find an equation of the line that is perpendicular to the line $2x+y = 2$ and containing the point $(-3,\ 0)$.

Problem I. A truck rental company rents a moving truck for one day by charging \$29 plus \$0.07 per mile. Write a linear equation that relates the cost C, in dollars, of renting the truck to the number x of miles driven. What is the cost of renting the truck if the truck is driven 110 miles?

Problem II. Suppose that the demand d (in thousands) for a certain mathematics text is related to the price P (in dollars) by the equation $d + 160p = 18,000$.

(1) Write the demand d in terms of the price p. Use your calculator to graph their relation.

(2) What is the demand when the price is 50?

(3) How will the demand change if the price is decreased from \$50 to \$46?

(4) How will the demand change if the price is increased from \$37.50 to \$38?

(5) At what price will there no longer be any demand for the text?

Chapter 8

Functions: Notation, Graphs, Domain, and Range

A function is a collection of points, $(x,\ y)$, such that no two points have the same x value and different y value. The domain of a function is its set of possible x values. The range of a function is its resulting set of possible y values.

Example 1:
a) Given: $\{(1,\ 1),\ (2,\ 4),\ (3,\ 8),\ (4,\ 9)\}$, this relation is a function because there are no ordered pairs with the same first coordinate and different second coordinate. The domain of this function is $\{1,\ 2,\ 3,\ 4\}$ and the range is $\{1,\ 4,\ 8,\ 9\}$.

b) Given: $\{(-2,\ 5),\ (-2,\ 4),\ (0,\ 0),\ (1,\ 3),\ (-2,\ 7)\}$, this relation is not a function because there are two ordered pairs, $(-2,\ 5)$ and $(-2,\ 7)$, that have same first coordinate, but different second coordinates.

Functions are often denoted by letters such as f, g, h, F, G, H, and others. The **vertical line test** is an easy way to determine whether a graph is a function. Imagine a vertical line placed anywhere along the graph. If the line intersects the graph in more than one place, the graph is not a function because two or more points will have the same x value.
For examples:

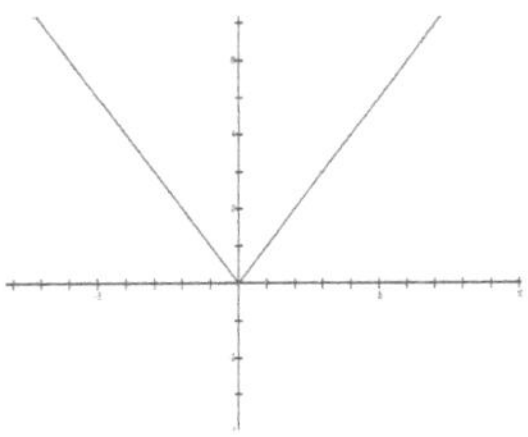

Function

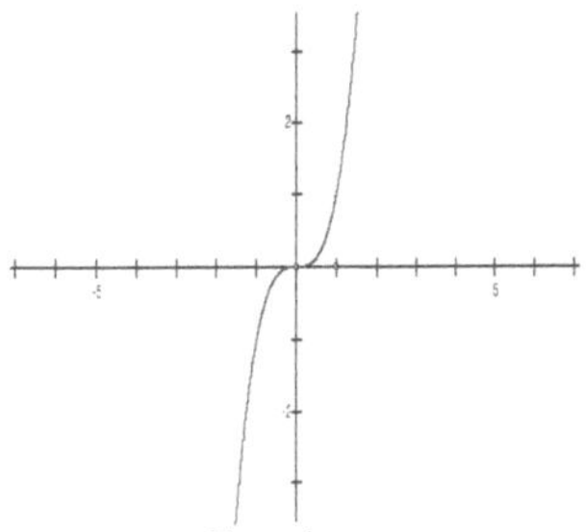

Function

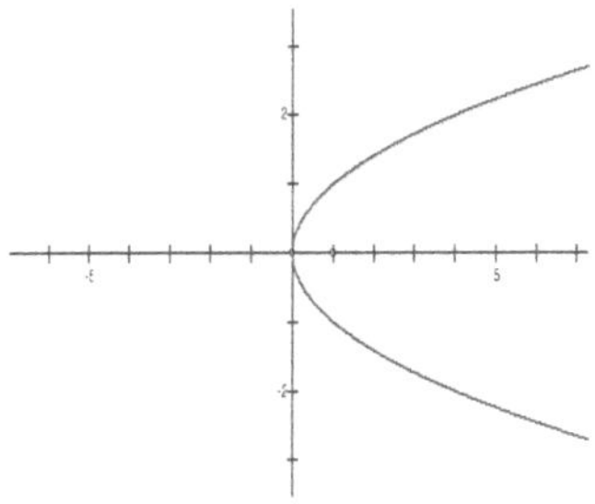

Not a function

Consider the following function $f(x) = 2x^3 - 4x^2 + x - 6$, we can find the value of this function at a give x value, for example:
a) $f(0) = -6$
b) $f(1) = -7$
c) $f(-1) = -13$
d) $f(-x) = -2x^3 - 4x^2 - x - 6$
e) $f(x+h) = 2(x+h)^3 - 4(x+h)^2 + x + h - 6$.

We can also determine whether a given equation is a function. For example: Determine whether the equation $x^2 + y^2 = 1$, which defines the unit circle, is a function, we need to solve the equation for y.

$$y^2 = 1 - x^2$$
$$y = \pm\sqrt{1 - x^2}$$

For any value of x between -1 and 1, two values of y result. This means that the equation does not define a function.

Rules for Determining the Domain:

(1) For the function $f(x) = \sqrt{x}$, $x \geq 0$, then the domain is $[0,\ \infty)$.

(2) For the function $f(x) = \dfrac{1}{x}$, $x \neq 0$, then the domain is $(-\infty,\ 0) \cup (0,\ \infty)$.

(3) For the function $f(x) = \dfrac{1}{\sqrt{x}}$, $x > 0$ (since $x \geq 0$ and $x \neq 0$), then the domain is $(0,\ \infty)$.

(4) Given $f(x) = |x|$, then the domain is all real numbers.

(5) All other functions will be studied later on.

Example 2: Determine the domain of the following functions:

(1) $g(x) = \sqrt{3x+2}$

$3x + 2 \geq 0$, so $x \geq \dfrac{-2}{3}$ or D= $[\dfrac{-2}{3},\ \infty)$.

(2) $h(x) = \dfrac{1}{2x+5}$

$2x + 5 \neq 0$, so $x \neq \dfrac{-5}{2}$ or D= $(-\infty,\ \dfrac{-5}{2}) \cup (\dfrac{-5}{2},\ \infty)$.

(3) $k(x) = \dfrac{1}{\sqrt{2x+3}}$

$2x + 5 > 0$, so $x > \dfrac{-5}{2}$ or D= $(\dfrac{-5}{2},\ \infty)$.

(4) $f(x) = \sqrt{x^2+4}$ (Think about it!)

Example 3: Given $f(x) = 3x^2$ and $g(x) = x^2 - 2x + 5$, find the following.

(1) $(f+g)(x) = 4x^2 - 2x + 5$

(2) $(f-g)(x) = 2x^2 + 2x - 5$

(3) $(f.g)(x) = 3x^4 - 6x^3 + 15x^2$

(4) $(\dfrac{f}{g})(x) = \dfrac{3x^2}{x^2 - 2x + 5}$

Example 4: Which of the following graphs represent a graph of a function.

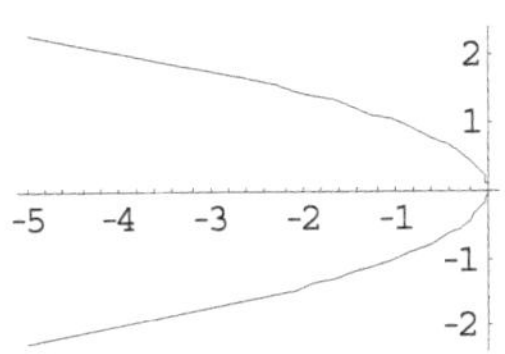

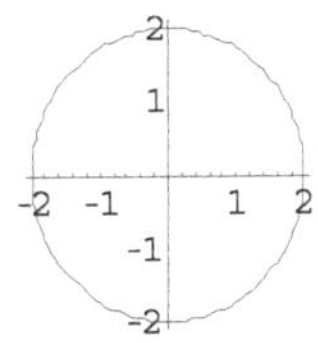

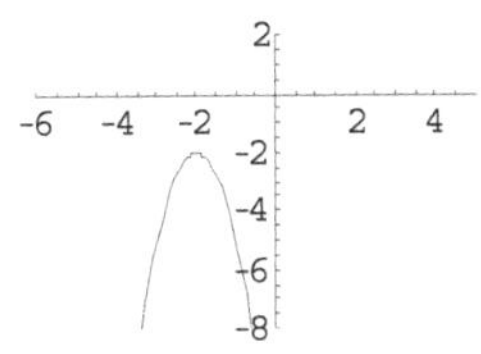

Only the last graph represent a graph of a function. The first two graphs are not.

Group Exercises 8

Functions: Notation, Graphs, Domain, and Range

(1) Determine whether each relation represents a function.

(a) $\{(2,\ 6),\ (-3,\ 6),\ (4,\ 9),\ (2,\ 10)\}$

(b) $\{(1,\ 2),\ (3,\ 4),\ (4,\ 6),\ (5,\ 4)\}$

(2) For $f(x) = 3x^2 + 2x - 4$, find each value:

(a) $f(0) =$

(b) $f(-1) =$

(c) $f(1) =$

(d) $f(-x) =$

(e) $f(x + h) =$

(3) Which of the following are functions?

(a)

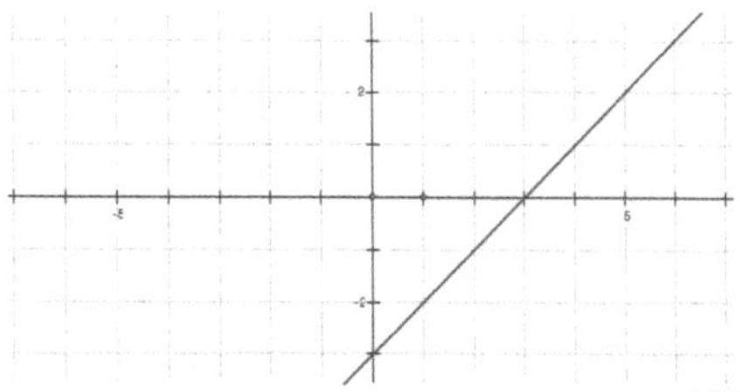

(b)

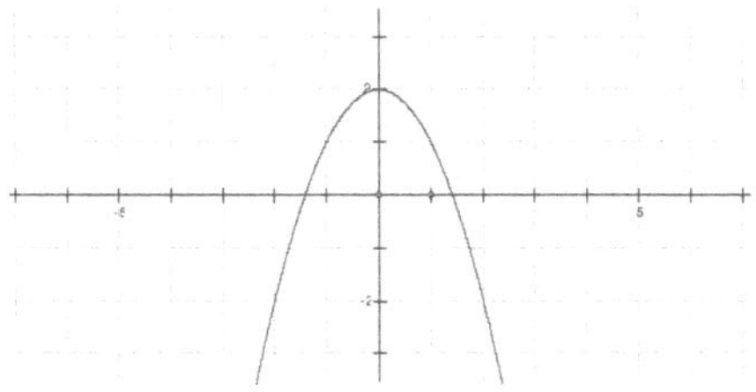

(c)

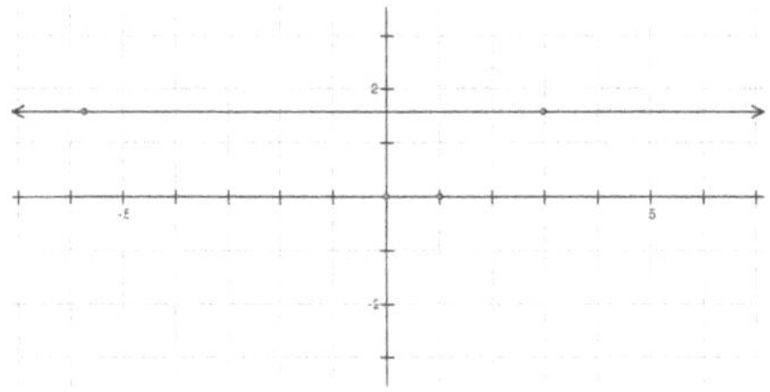

(4) Determine whether the equation is a function.

(a) $y = x^3$

(b) $y^2 = 4 - x^2$

(c) $x = y^2$

(d) $y = |x|$

(5) Find the domain of each function. Write your answer using interval notation.

(a) $y = \sqrt{3 - 2x}$

(b) $g(x) = \dfrac{x}{x^2 - 16}$

(c) $h(t) = \dfrac{t}{\sqrt{t+5}}$

(d) $f(x) = \dfrac{1}{x} - \dfrac{x+1}{x-4}$

(6) State the domain and the range of each of the following. Write your answer using interval notation.

(a)

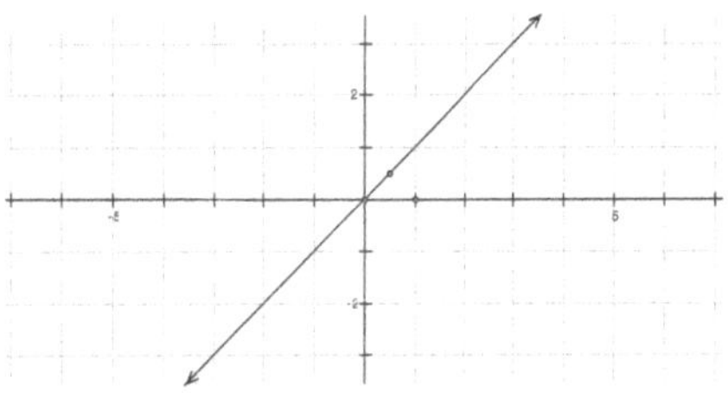

(b)

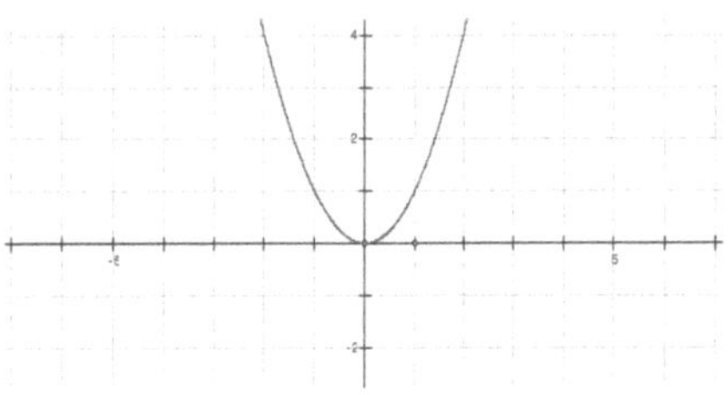

(c)

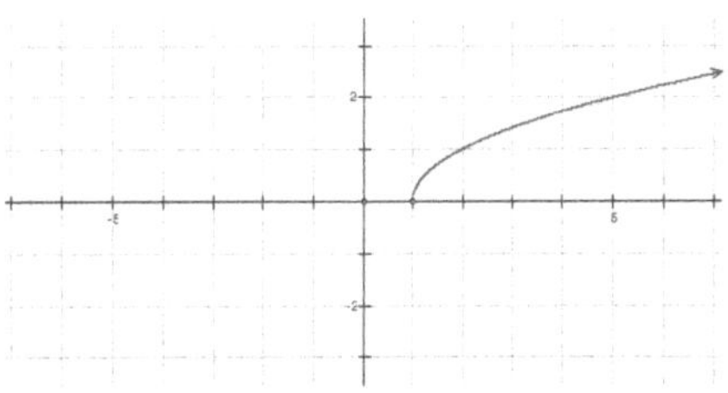

(7) For the given functions $f(x) = 3x + 4$ and $g(x) = 2x - 3$, find the following functions:

(a) $f + g$

(b) $f - g$

(c) $f \cdot g$

(d) $\frac{f}{g}$

Practice Exercises 8

Functions: Notation, Graphs, Domain, and Range

(1) Determine whether each relation represents a function.

(a) $\{(1,\ 1),\ (4,\ 2),\ (3,\ 1),\ (4,\ -2)\}$

(b) $\{(0,\ 2),\ (3,\ -1),\ (1,\ 3),\ (8,\ -1)\}$

(2) For $g(x) = -|x - 3|$, find each value:

(a) $g(0) =$

(b) $g(-3) =$

(c) $g(3) =$

(d) $g(-x) =$

(e) $g(x + h) =$

(3) Which of the following are functions?

(a)

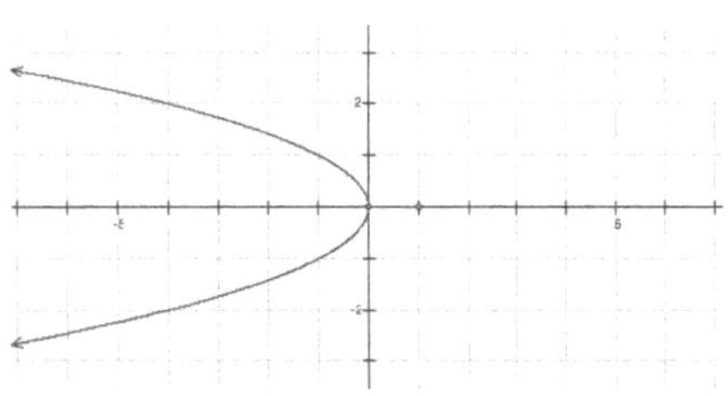

(b)

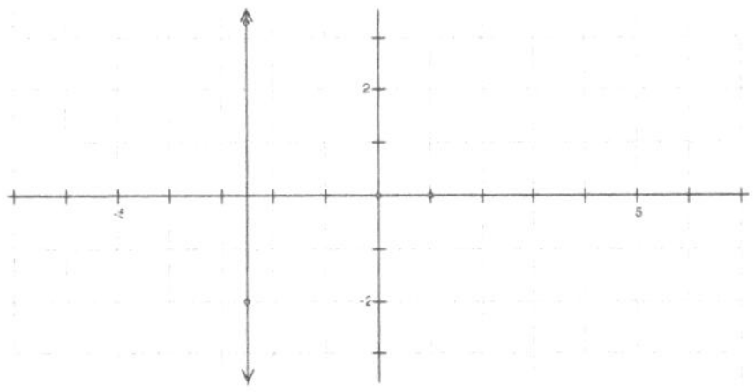

(c)

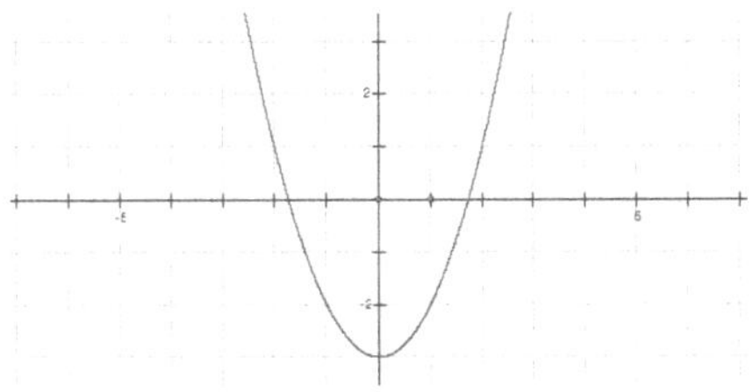

(4) Determine whether the equation is a function.

(a) $y = x^2$

(b) $y = \pm\sqrt{1 - 2x}$

(c) $x = 1 - y^2$

(d) $y = |x + 1|$

(5) Find the domain of each function. Write your answer using interval notation.

(a) $f(x) = \sqrt{-x - 2}$

(b) $y = \dfrac{4}{\sqrt{x-9}}$

(c) $g(t) = \dfrac{t+4}{\sqrt{t^3-4t}}$

(d) $h(x) = -2|x-1| - 3$

(6) State the domain and the range of each of the following. Write your answer using interval notation.

(a)

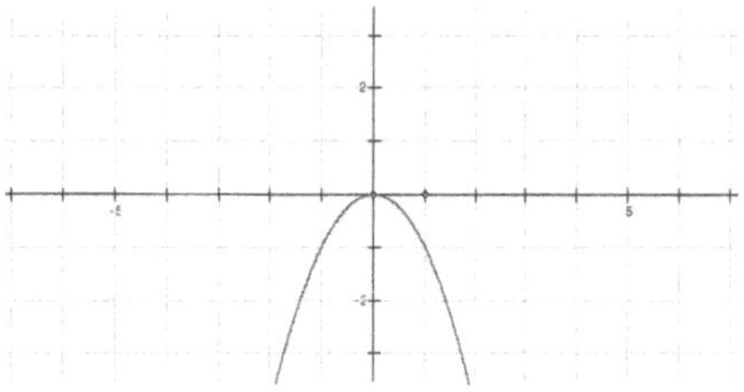

(b)

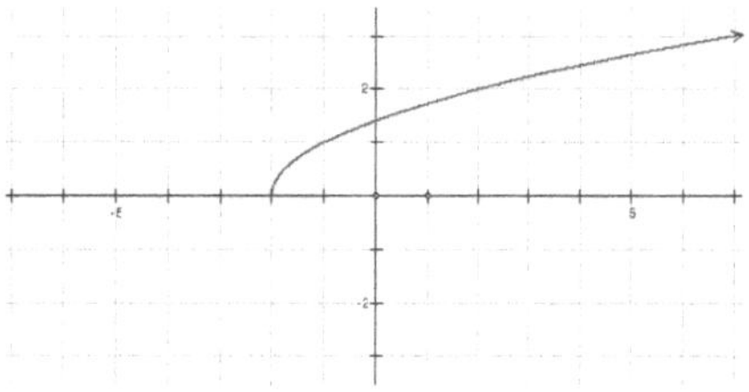

(c)

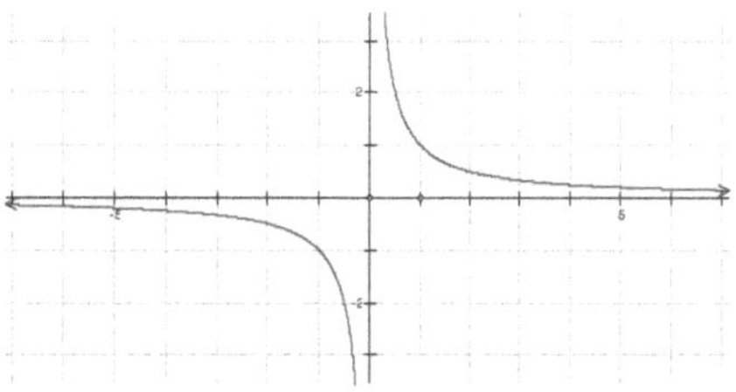

(7) For the given functions $f(x) = \sqrt{x}$ and $g(x) = 3x - 5$, find the following functions:

(a) $f + g$

(b) $f - g$

(c) $f \cdot g$

(d) $\dfrac{f}{g}$

Chapter 9

Graphs of Functions

Example 1: Consider the function $f(x) = \dfrac{x+2}{x-6}$, answer the following questions:

(1) Is the point $(2,\ 1)$ on the graph of f?

We check $f(2)$, if it is 1, then the point is on the graph of f. But $f(2) = -1$, so the point $(2,\ 1)$ is not on the graph of f.

(2) If $x = 4$, what is $f(x)$? What point is on the graph of f?

$f(4) = \dfrac{4+2}{4-6} = \dfrac{6}{-2} = -3$. So the point $(4,\ -3)$ is on the graph of f.

(3) If $f(x) = 3$, what is x? What point(s) are on the graph of f?

Solve the equation $\dfrac{x+2}{x-6} = 3$ for x. Then we have $x = 10$, and therefore, the point $(10,\ 3)$ is on the graph of f.

(4) List the x-intercept, if any, of the graph of f.

To find the x-intercept, we let $y = 0$ and solve the equation $\dfrac{x+2}{x-6} = 0$ for x. Then $x = -2$.

(5) List the $y-$intercept, if there is one, of the graph of f.

To find the $y-$intercept, we let $x = 0$ and solve the equation $y = \dfrac{0+2}{0-6}$ for y. So $y = \dfrac{-1}{3}$

"How We Obtain Information from the Graph of a Function"

Consider the graph of f below

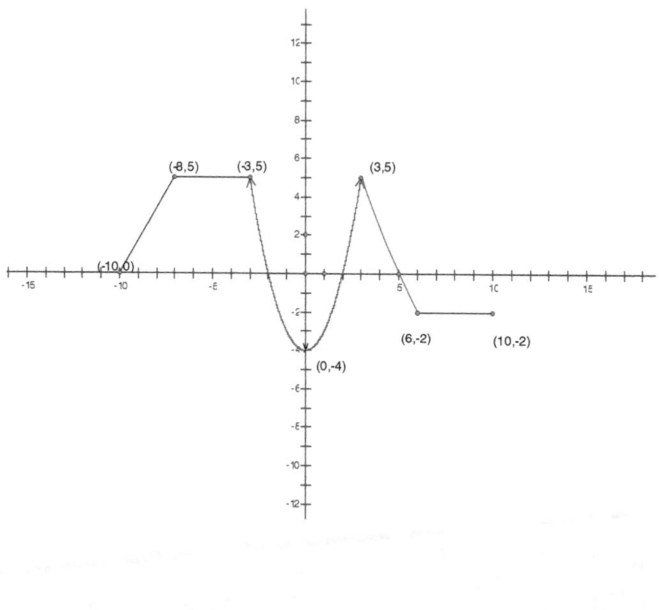

(1) Find $f(-8)$, $f(0)$, $f(10)$, and $f(-10)$.

$f(-8) = 5$, $f(0) = -4$, $f(10) = -2$, and $f(-10) = 0$.

(2) For what numbers x is $f(x) = 0$? $x = -10, -2, 2, 5$.

(3) What are the $x-$intercepts? $x = -10, -2, 2, 5$.

(4) What is the $y-$intercept? $y = -4$

(5) Is $f(8)$ positive or negative? negative.

(6) Is $f(-5)$ positive or negative? positive.

"Determining Even or Odd Functions from the Graph"

If the graph is symmetric with respect to the $y-$axis, then the function is even, and if the graph is symmetric with respect to the origin, then the function is odd.

Example 2: Which of the following graphs is even, odd, or neither.

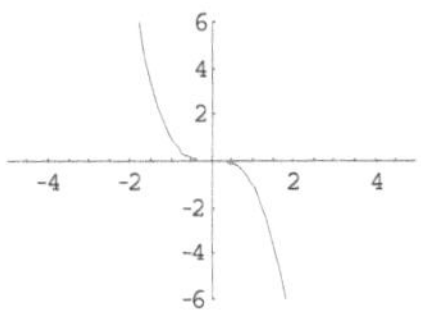

odd

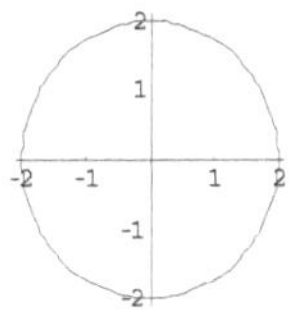

neither

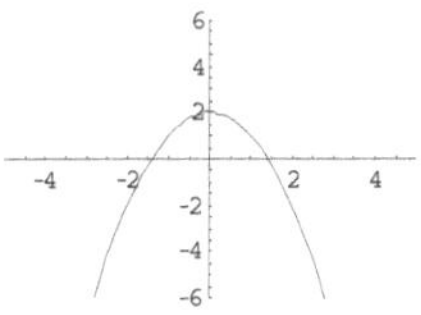

even

A function f is **even** if, for every number x in its domain, the number $-x$ is also in the domain and $f(-x) = f(x)$.

A function f is **odd** if, for every number x in its domain, the number $-x$ is also in the domain and $f(-x) = -f(x)$.

Example 3 : Determine algebraically whether each of the following functions is even, odd, or neither.

(1) $f(x) = x^2 - 4$

$f(-x) = (-x)^2 - 4 = x^2 - 4 = f(x)$, so f is even function.

(2) $g(x) = x^3 - 2$

$g(-x) = (-x)^3 - 2 = -x^3 - 2 = -(x^3 + 2)$, which is neither $g(x)$ nor $-g(x)$, so g is neither odd nor even.

(3) $h(x) = |x|$

$h(-x) = |-x| = |x| = h(x)$, so h is even function.

(4) $F(x) = 5x^3 - x$

$F(-x) = 5(-x)^3 - (-x) = -5x^3 + x = -(5x^3 - x) = -F(x)$, so F is odd function.

"Average Rate of Change"

The average rate of change of f from x_1 to x_2 is

$$\frac{\Delta y}{\Delta x} = \frac{f(x_2) - f(x_1)}{x_2 - x_1}, \quad x_2 \neq x_1.$$

This is also called a difference quotient.

Example 4: Find the average rate of change of $f(x) = 3x^2 - 2x + 1$

(1) from 0 to 1

$$\frac{f(x_2) - f(x_1)}{x_2 - x_1} = \frac{f(1) - f(0)}{1 - 0} = \frac{2 - 1}{1} = 1$$

(2) from 1 to 3

$$\frac{f(x_2) - f(x_1)}{x_2 - x_1} = \frac{f(3) - f(1)}{3 - 1} = \frac{22 - 2}{2} = 10$$

(3) from -2 to 0

$$\frac{f(x_2) - f(x_1)}{x_2 - x_1} = \frac{f(0) - f(-2)}{0 - (-2)} = \frac{1 - 17}{2} = -8$$

Group Exercises 9

Graphs of Functions

(1) Given the function $f(x) = \dfrac{x+2}{x-6}$

(a) Is the point $(-1,\ 2)$ on the graph?

(b) If $x = -2$, what is $f(x)$?

(c) If $f(x) = -1$, what is x?

(2) Use the graph

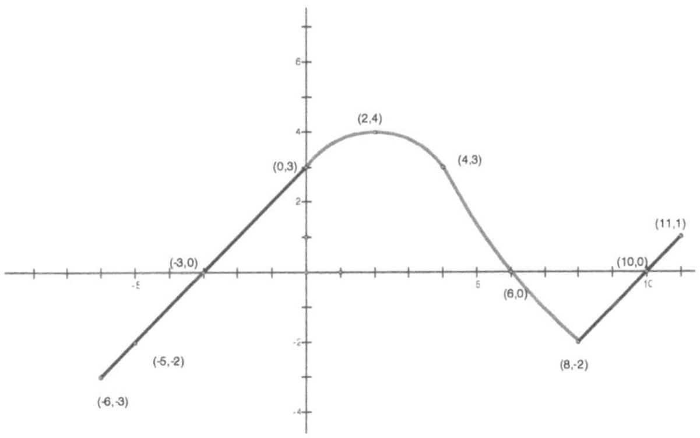

(a) Find $f(0)$, $f(-6)$, $f(6)$ and $f(11)$.

(b) For what number x is $f(x) = 0$?

(c) What are the x-intercepts?

(d) What is the y-intercept?

(e) Is $f(-4)$ positive or negative?

(f) Is $f(3)$ positive or negative?

(3) Determine algebraically whether each function is even, odd, or neither.

(a) $f(x) = 4x^3$

(b) $h(x) = \dfrac{x}{x^2 - 1}$

(c) $g(x) = \sqrt{x}$

(d) $F(x) = x + |x|$

(4) Compute the average rate of change of $f(x) = -2x^2 + 4$.

(a) From 0 to 2

(b) From 1 to 3

(c) From 1 to 4

Practice Exercises 9

Graphs of Functions

(1) Given the function $g(x) = \dfrac{x^2+2}{x+4}$

(a) Is the point $(-1,\ 2)$ on the graph?

(b) If $x = 0$, what is $g(x)$?

(c) If $g(x) = \frac{1}{2}$, what is x? What point(s) are on the graph of g?

(2) Use the graph

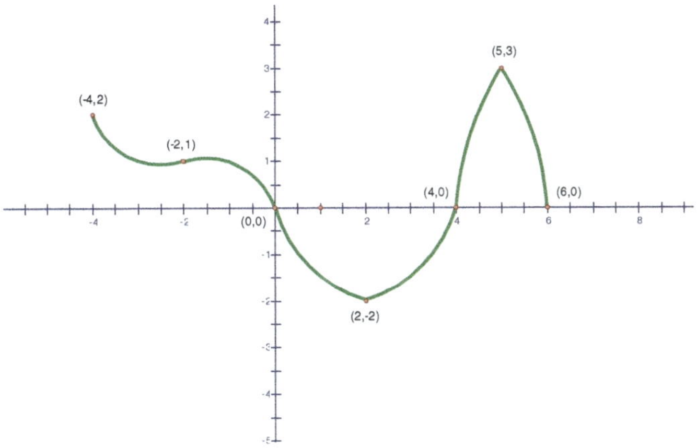

(a) Find $f(0)$, $f(-2)$, $f(6)$ and $f(5)$.

(b) For what number x is $f(x) = 0$?

(c) What are the x-intercepts?

(d) What is the y-intercept?

(e) Is $f(-4)$ positive or negative?

(f) Is $f(3)$ positive or negative?

(3) Determine algebraically whether each function is even, odd, or neither.

(a) $f(x) = 3x^3 + 5$

(b) $h(x) = \dfrac{-x^3}{3x^2 - 9}$

(c) $g(x) = \dfrac{1}{x^2}$

(d) $F(x) = \dfrac{x}{|x|}$

(4) Compute the average rate of change of $f(x) = -x^3 + 1$

(a) From 0 to 2

(b) From 1 to 3

(c) From -1 to 1

Chapter 10

Library of Functions

Constant Function $f(x) = b$ (Horizontal Line)

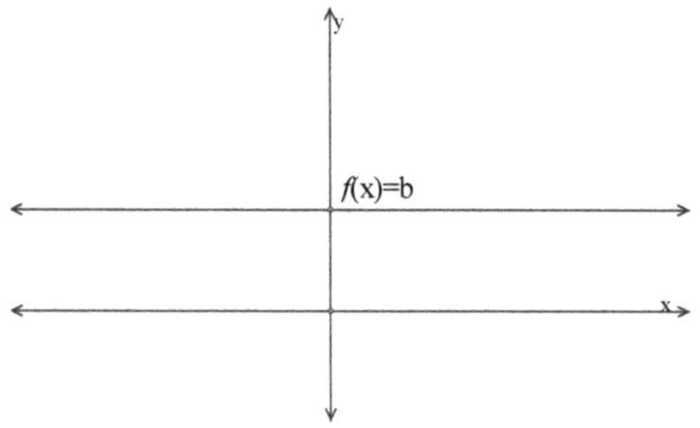

Linear Function $f(x) = mx + b$ (Straight Line)

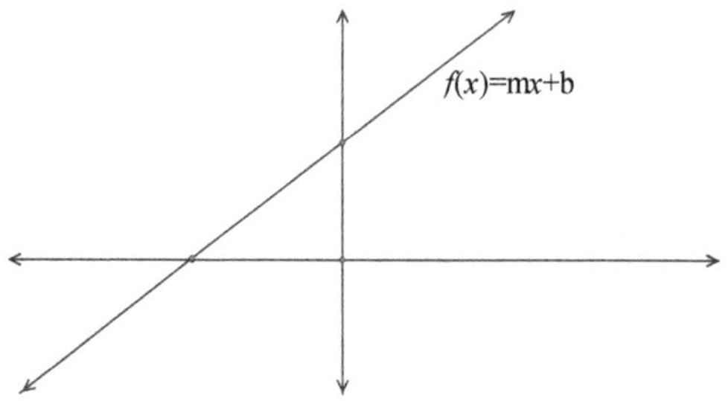

Identity Function $f(x) = x$ (Straight Line)

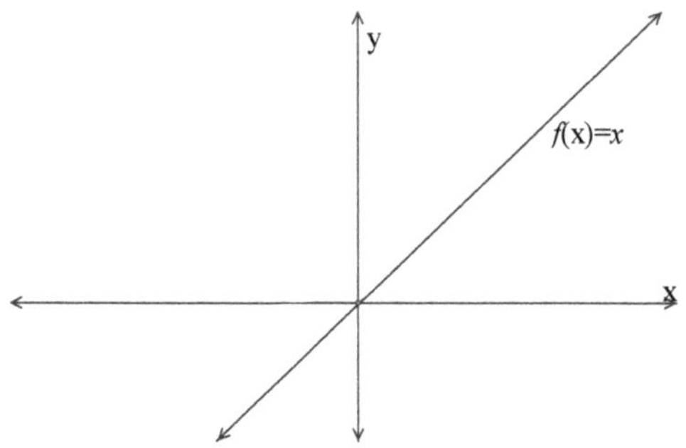

Absolute Value Function $f(x) = |x| = \begin{cases} x & \text{if } x \geq 0 \\ -x & x < 0 \end{cases}$

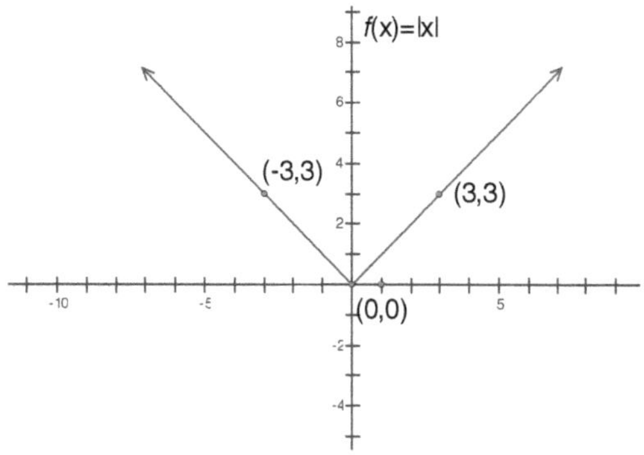

Square Root Function $f(x) = \sqrt{x}$

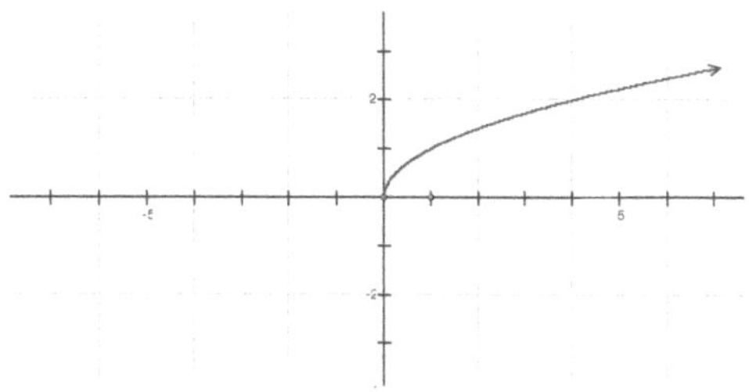

Square Function $f(x) = x^2$

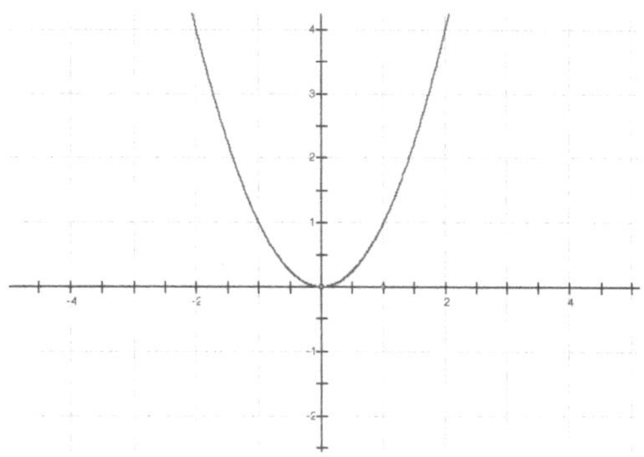

Cube Function $f(x) = x^3$

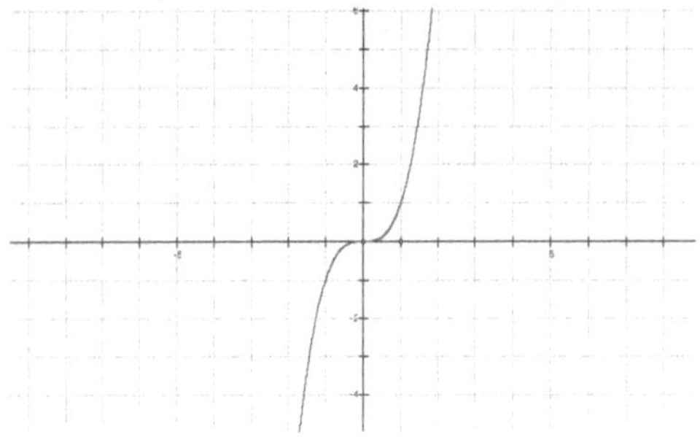

Reciprocal Function $f(x) = \frac{1}{x}$

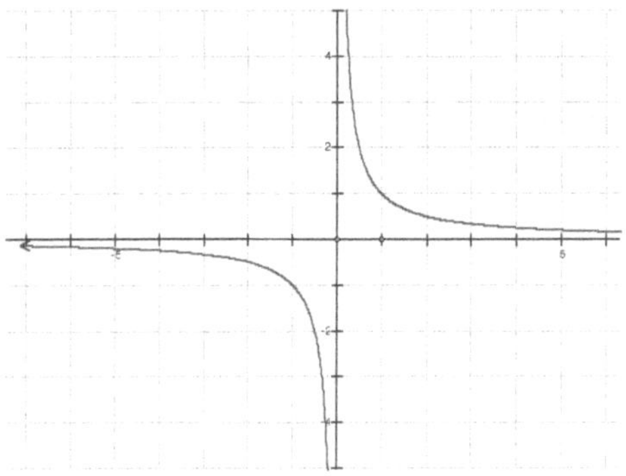

Piecewise-defined Functions

The absolute value function defined above is an example of a piecewise-defined function, i.e., a function that is defined by more than one equation.

Example : For $f(x)$, (a) find the domain, (b) locate the intercepts, (c) graph the function, and then (d) find the range.

$$f(x) = \begin{cases} x+3 & \text{if } -3 \leq x < 0 \\ 3 & \text{if } x = 0 \\ \sqrt{x} & x > 0 \end{cases}$$

The domain is $(-3,\ \infty)$ and the range is $(0,\ \infty)$.
The $x-$intercept is $x = -3$ and the $y-$intercept is $y = 3$.

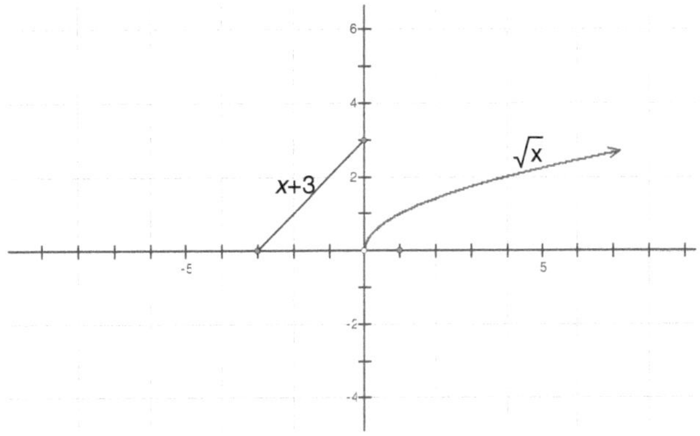

Problem I. Holders of credit cards issued by banks, department stores, oil companies, and so on, receive bills each month that state minimum amounts that must be paid by a certain due date. The minimum due depends on the total amount owed. One such credit card company uses the following rules: For a bill of less than \$10, the entire amount is due. For a bill of at least \$10 but less than \$500, the minimum due is \$10. There is a minimum of \$30 due on a bill of at least \$500 but less than \$1000, a minimum of \$50 due on a bill of at least \$1000 but less than \$1500, and a minimum of \$70 due on bills of \$1500 or more. Find the function m that describes the minimum payment due on a bill of x dollars.

Solution:

$$m(x) = \begin{cases} x & \text{if } 0 \leq x < 10 \\ 10 & \text{if } 10 \leq x < 500 \\ 30 & \text{if } 500 \leq x < 1000 \\ 50 & \text{if } 1000 \leq x < 1500 \\ 70 & \text{if } x \geq 1500 \end{cases}$$

Sketch the graph of m (exercise!)

Group Exercises 10

Library of Functions

(1) Sketch the graph of each function and then find the domain and range.

(a) $f(x) = 3x + 1$

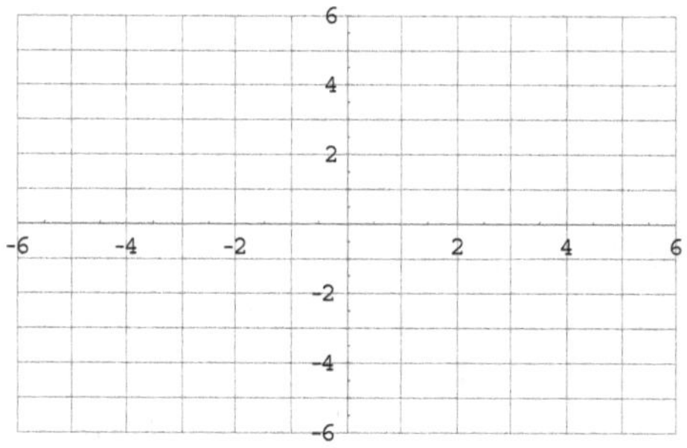

(b) $f(x) = 4 - x^2$

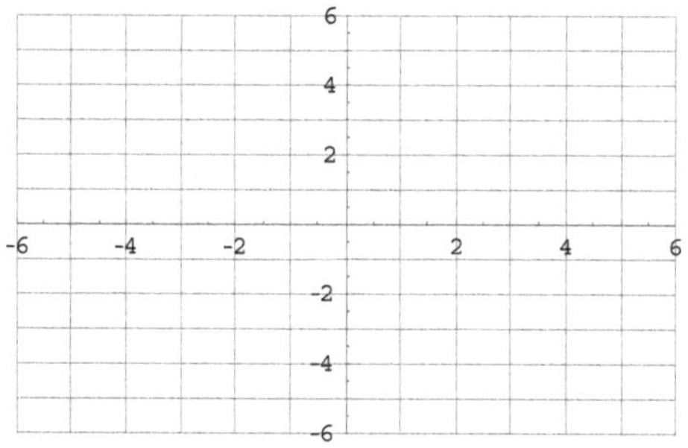

(c) $f(x) = |2x + 1|$

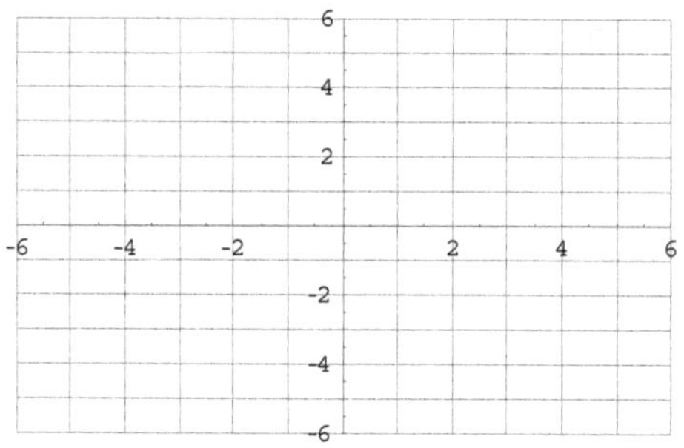

(d) $f(x) = \begin{cases} x & \text{if } x < 2 \\ 2 & \text{if } x \geq 2 \end{cases}$

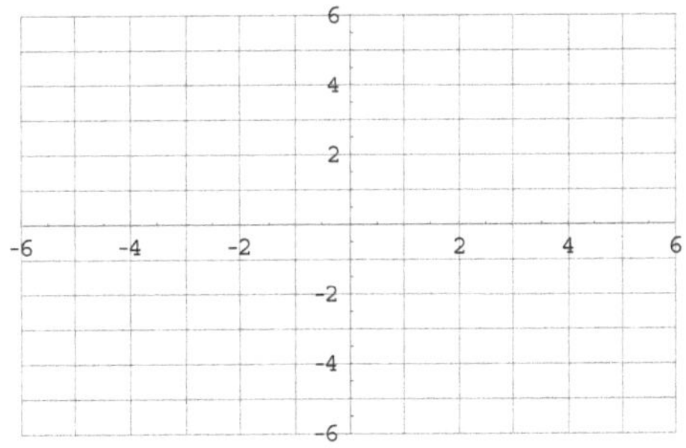

(2) If $g(x) = \begin{cases} x^2 & \text{if } x < 0 \\ 2 & \text{if } x = 0 \\ 2x + 1 & \text{if } x > 0 \end{cases}$

Find:

(a) $g(-2) =$

(b) $g(0) =$

(c) $g(2) =$

(d) the domain of $g(x)$.

(e) the range of $g(x)$.

Problem I. Sprint PCS offers a monthly cellular phone plan for \$39.99. It includes 350 anytime minutes plus \$0.25 per minute for additional minutes. The following function is used to compute the cost for a subscriber:

$$C(x) = \begin{cases} 39.99 & \text{if } 0 < x \leq 350 \\ 0.25x - 47.51 & \text{if } x > 350 \end{cases}$$

where x is the number of anytime minutes used. Compute the monthly cost of the cellular phone for use of the following anytime minutes:

(a) 200

(b) 365

(c) 351

Practice Exercises 10

Library of Functions

(1) Sketch the graph of each function and then find the domain and range.

(a) $f(x) = 3x - 2$

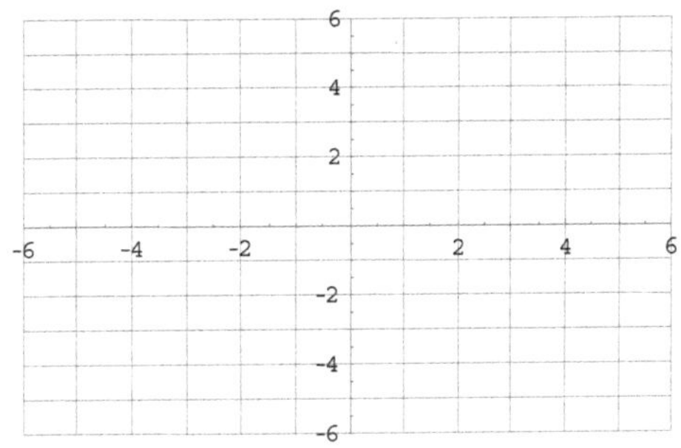

(b) $f(x) = x^2 + 4$

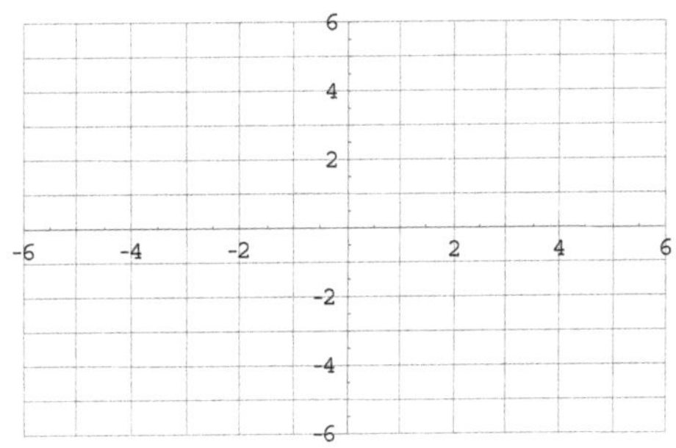

(c) $f(x) = |2x - 1|$

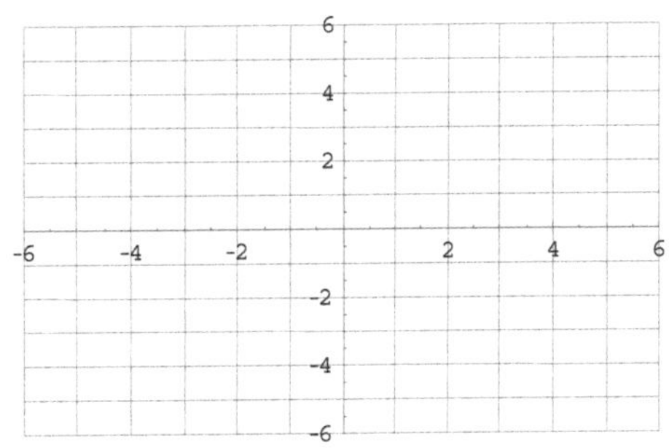

(d) $f(x) = \begin{cases} 3+x & \text{if } -4 \leq x < 0 \\ 3 & \text{if } x = 0 \\ \sqrt{x} & \text{if } x > 0 \end{cases}$

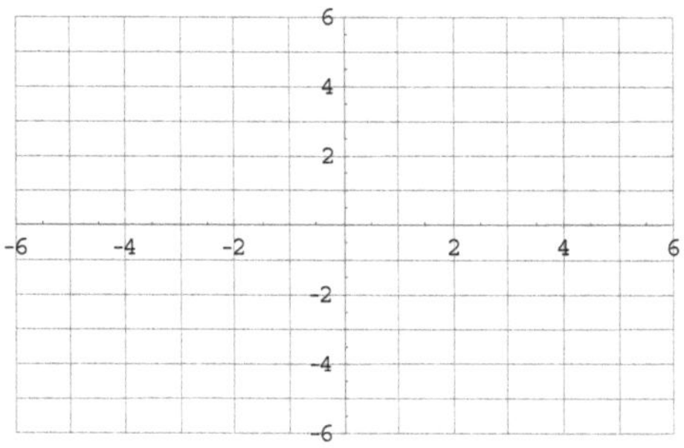

(2) If $h(x) = \begin{cases} x^3 & \text{if } x < 0 \\ 3x+2 & \text{if } x \geq 0 \end{cases}$

Find:

(a) $h(-1) =$

(b) $h(0) =$

(c) $h(1) =$

(d) the domain of $h(x)$.

(e) the range of $h(x)$.

Problem I. According to the U.S. Postal Service, first-class mail is used for personal and business correspondence. Any mailable item may be sent as first-class mail. It includes postcards, letters, large envelopes, and small packages. The maximum weight is 13 ounces. The following function is used to compute the cost of mailing a first-class letter.

$$C(x) = \begin{cases} 0.39 & \text{if } 0 < x \leq 1 \\ 0.23 \operatorname{int}(x) + 0.37 & \text{if } 1 < x \leq 13 \end{cases}$$

where x is the weight of package in ounces. Compute the cost of mailing a package via first class for the following weights:

(a) A letter weighting 4.3 ounces

(b) A postcard weighting 0.4 ounce

(c) A package weighting 12.2 ounces

Chapter 11

Graphing Techniques and Transformations

Vertical Shift

If $c > 0$, the graph of $y = f(x) + c$ is the graph of $y = f(x)$ shifted up c units. Similarly, the graph of $y = f(x) - c$ is the graph of $y = f(x)$ shifted down c units.

Example 1: The following example illustrates the concepts of vertical shifts.
Consider the graph of $y = f(x) = x^2$, $y = x^2 + 2$, and $y = x^2 - 2$. The graph of $y = x^2 + 2$ is the graph of $y = x^2$ shifted up 2 units and the graph of $y = x^2 - 2$ is the graph of $y = x^2$ shifted down 2 units.

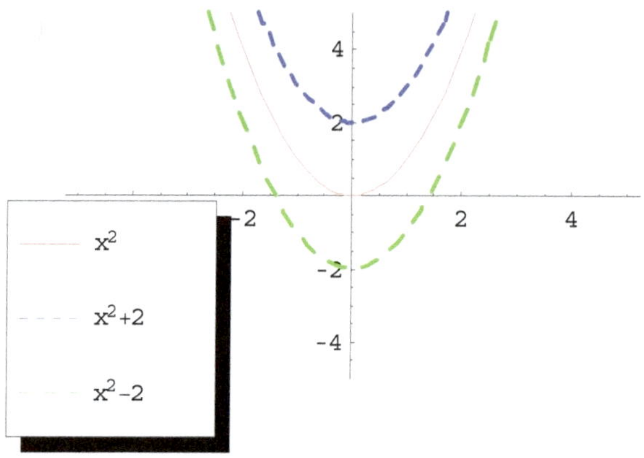

Horizontal Shift

If $c > 0$, the graph of $y = f(x - c)$ is the graph of $y = f(x)$ shifted c units to the right. Similarly, the graph of $y = f(x + c)$ is the graph of $y = f(x)$ shifted c units to the left.

Example 2: The following example illustrates the concepts of horizontal shifts.
Consider the graph of $y = f(x) = x^3$, $y = (x - 2)^3$, and $y = (x + 2)^3$. The graph of $y = (x - 2)^3$ is the graph of $y = x^3$ shifted 2 units to the right and the graph of $y = (x + 2)^3$ is the graph of $y = x^3$ shifted 2 units to the left .

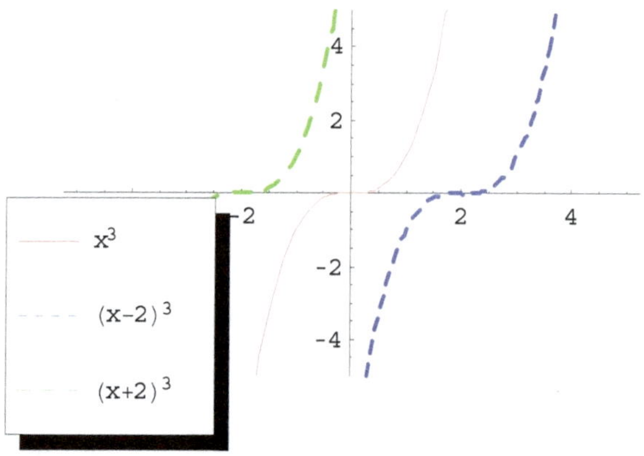

Compression and Stretches

If $a > 1$, the graph of $y = af(x)$ is the graph of $y = f(x)$ stretched vertically a units and if $0 < a < 1$, the graph of $y = af(x)$ is the graph of $y = f(x)$ compressed vertically a units. The vertical stretch and vertical compression can be obtained by multiplying each y-coordinate of $y = f(x)$ by a.

Example 3: The following example illustrates the concepts of vertical stretch and vertical compression.
Consider the graph of $y = f(x) = |x|$, $y = 2|x|$, and $y = \frac{1}{2}|x|$. The graph of $y = 2|x|$ is the graph of $y = |x|$ stretched vertically 2 units and the graph of $y = \frac{1}{2}|x|$ is the graph of $y = |x|$ compressed vertically $\frac{1}{2}$ unit.

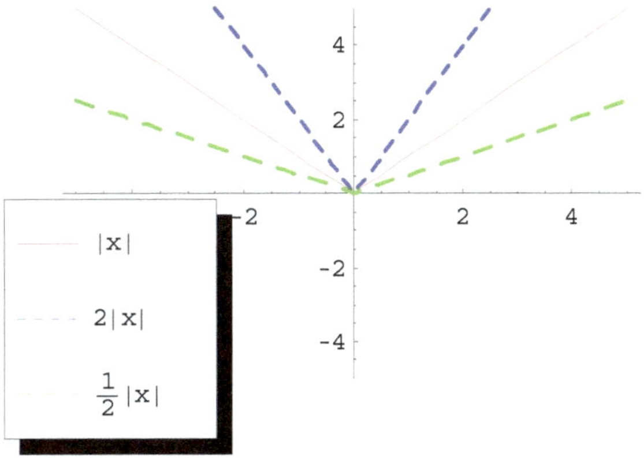

Reflections about the x-axis and the y-axis

The graph of $y = -f(x)$ is the reflection about the x-axis of the graph $y = f(x)$. The graph of $y = f(-x)$ is the reflection about the y-axis of the graph $y = f(x)$.

Example 4: The following example illustrates the concepts of reflection about the $x-$ and $y-$axis.
Consider the graph of $y = f(x) = \sqrt{x}$, $y = -\sqrt{x}$, and $y = \sqrt{-x}$. The graph of $y = -\sqrt{x}$ is the graph of $y = \sqrt{x}$ reflected about the $x-$axis and the graph of $y = \sqrt{-x}$ is the graph of $y = \sqrt{x}$ reflected about the $y-$axis.

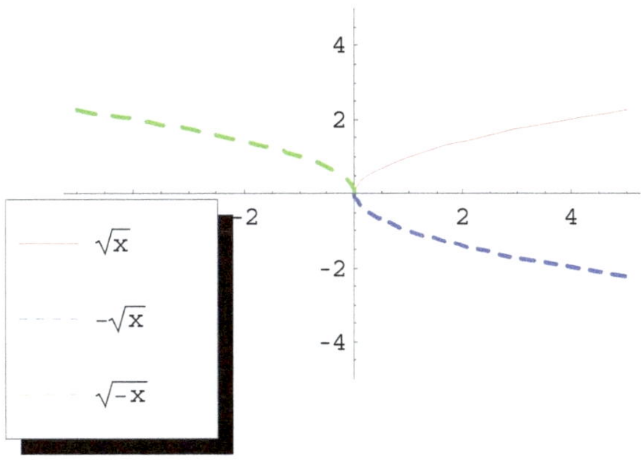

Group Exercises 11

Graphing Techniques and Transformations

(1) Graph each function using the techniques of shifting, compressing, stretching, and/or reflecting. Start with the basic function and show all stages.

(a) $f(x) = x^2 - 2$

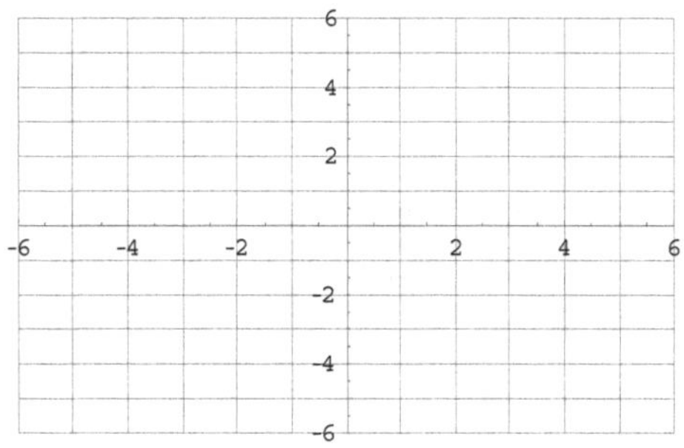

(b) $f(x) = x^3 - 1$

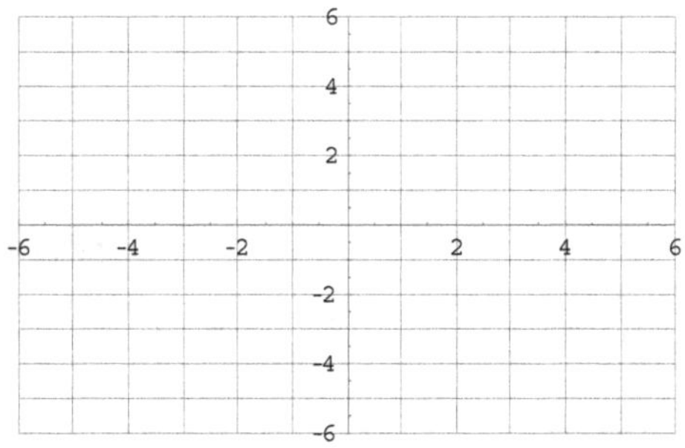

(c) $g(x) = \sqrt{x + 2}$

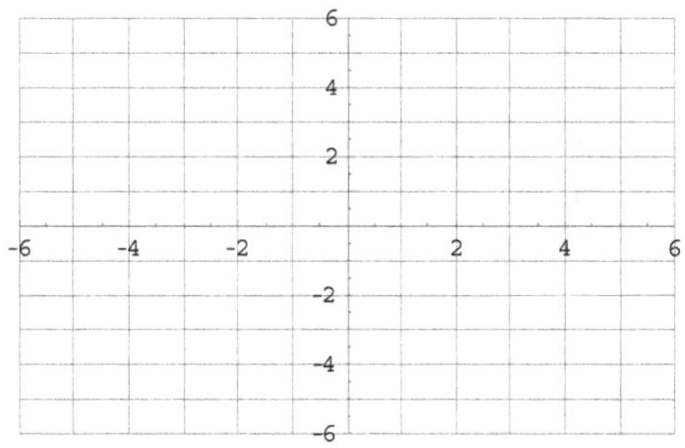

(d) $h(x) = 3(x-1)^2 + 1$

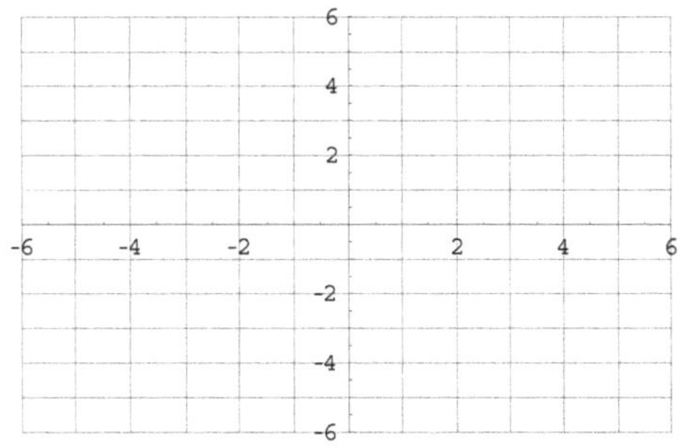

(e) $h(x) = -|x+1| + 3$

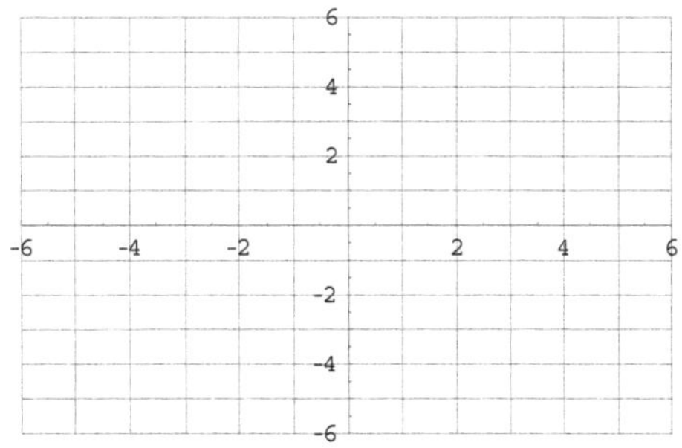

(f) $f(x) = \dfrac{1}{-x} + 2$

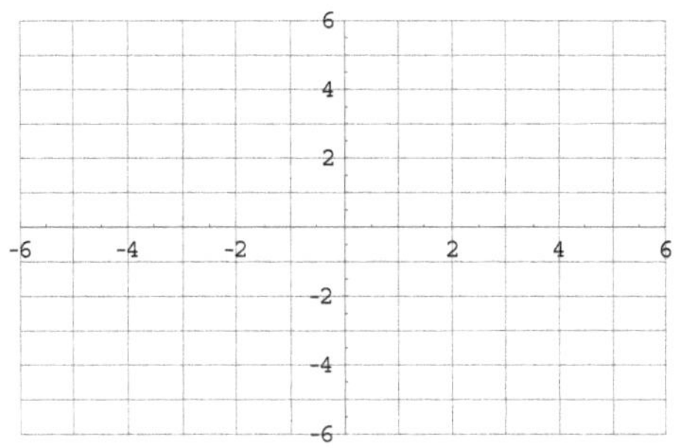

(g) $g(x) = \dfrac{1}{2}x^2 - 2$

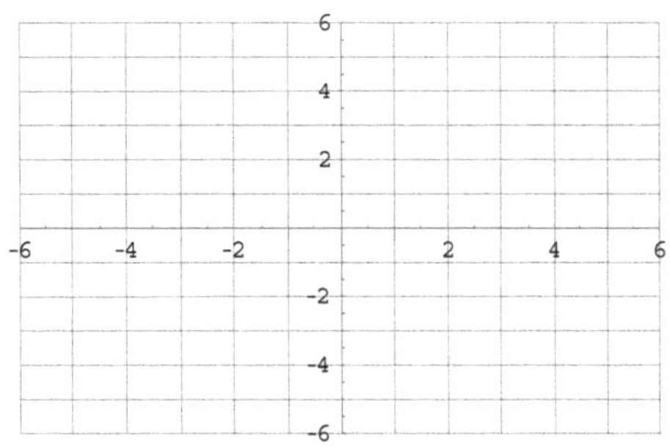

(2) Find the function that is finally graphed after the following transformations are applied to the graph of $y = \sqrt{x}$

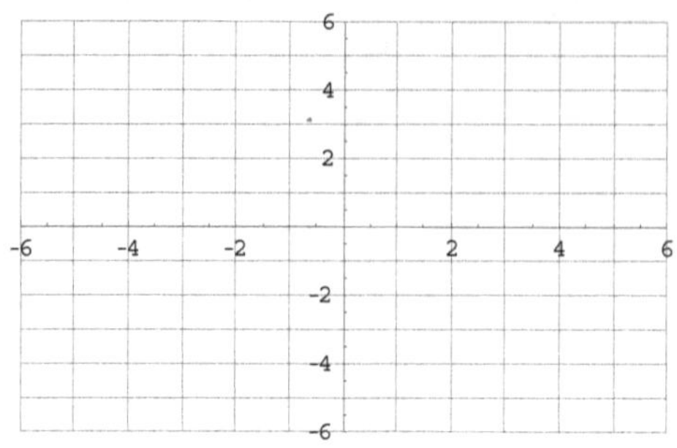

(a) Shift up 2 units

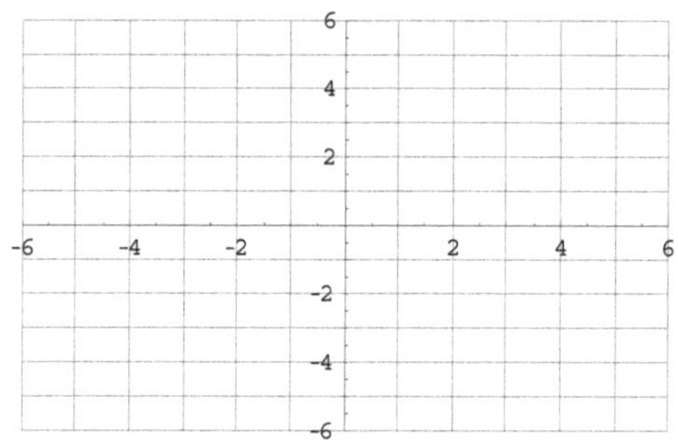

(b) Reflect about the x-axis

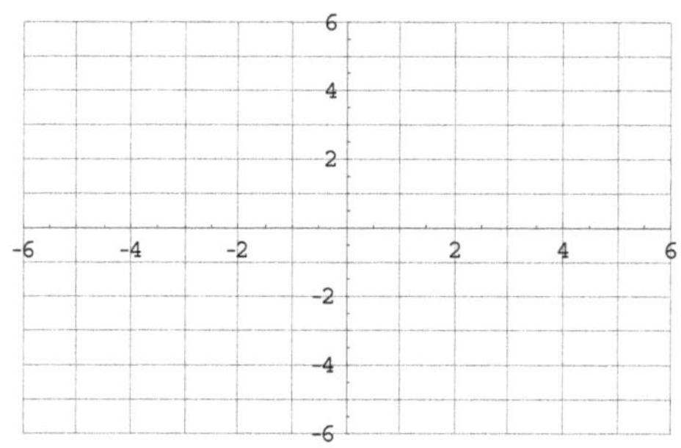

(c) Reflect about the y-axis

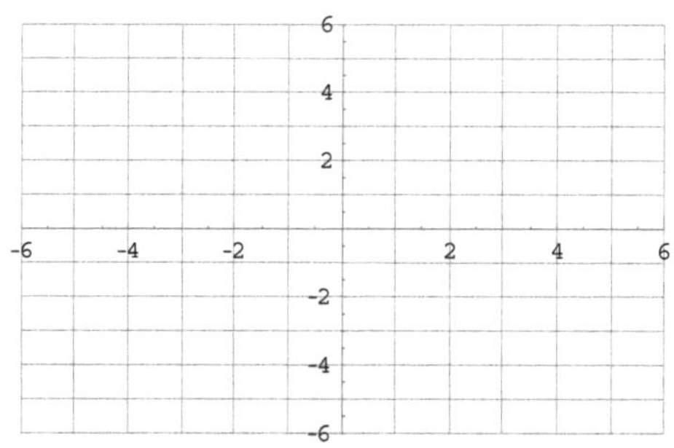

(3) The price p and the quantity x sold of a certain product obey the demand equation

$$p = -\frac{1}{6}x + 100, \qquad 0 \leq x \leq 600$$

(a) Express the revenue R as a function of x. (Recall: $R = x.p$)

(b) What is the revenue if 200 units are sold?

(c) Graph the revenue function.

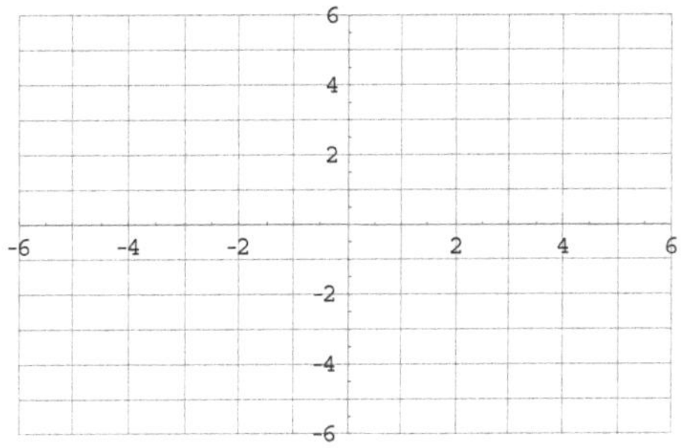

(d) What quantity x maximizes revenue? What is the maximum revenue?

Practice Exercises 11

Graphing Techniques and Transformations

(1) Graph each function using the techniques of shifting, compressing, stretching, and/or reflecting. Start with the basic function and show all stages.

(a) $f(x) = x^2 + 2$

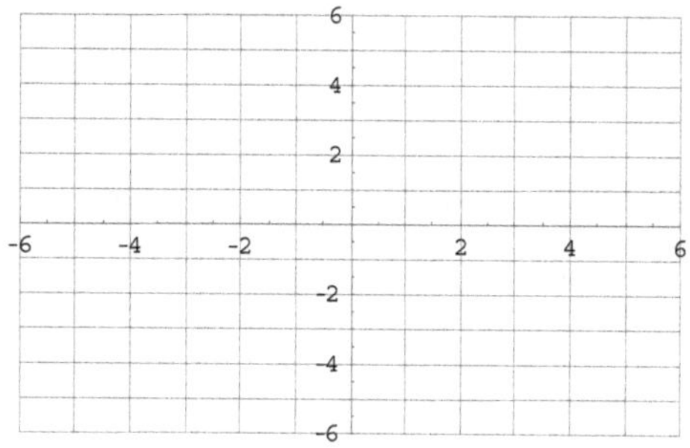

(b) $g(x) = x^3 + 1$

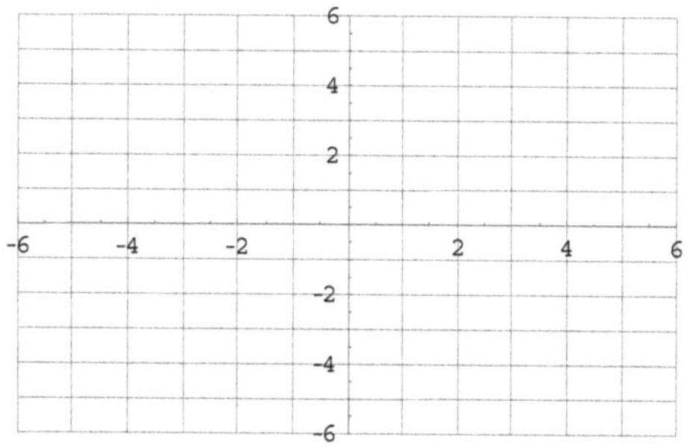

(c) $h(x) = \sqrt{x - 1}$

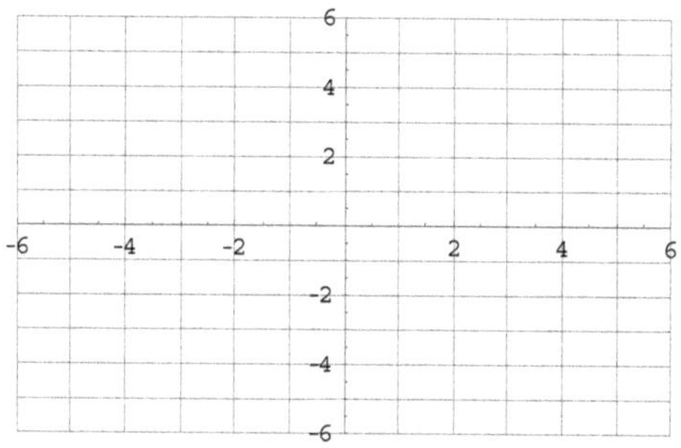

(d) $f(x) = 2(x+2)^2 - 1$

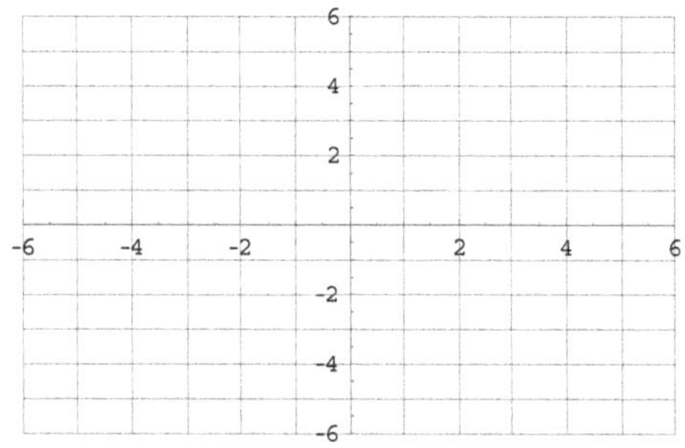

(e) $g(x) = |x-1| - 1$

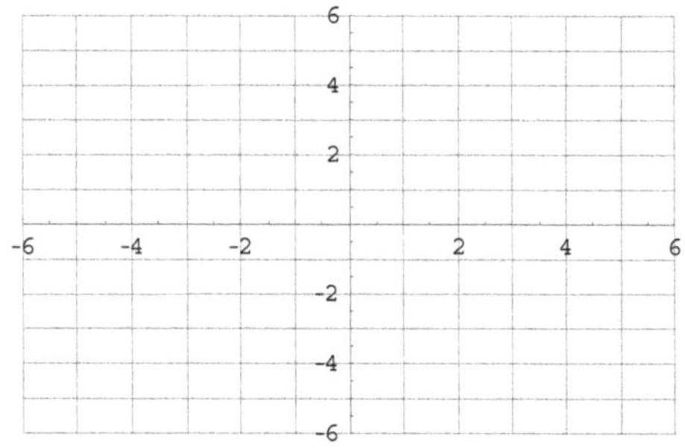

(f) $f(x) = \dfrac{1}{-x} + 2$

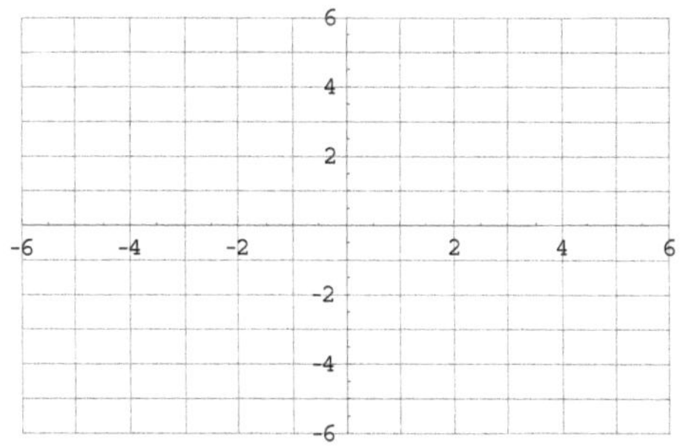

(g) $g(x) = \dfrac{-1}{2}x^2 + 1$

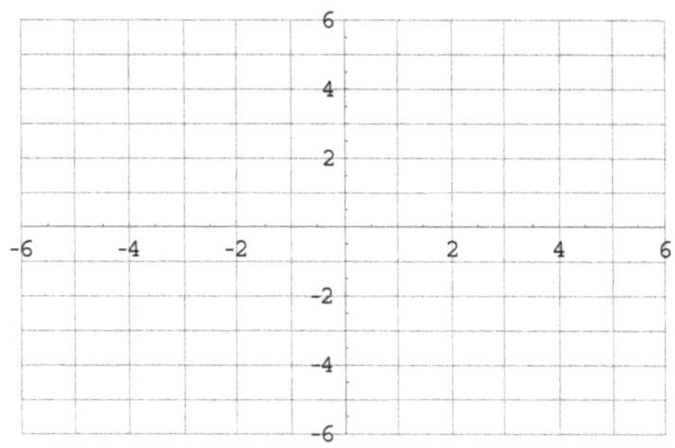

(2) Find the function that is finally graphed after the following transformations are applied to the graph of $y = \sqrt{x}$

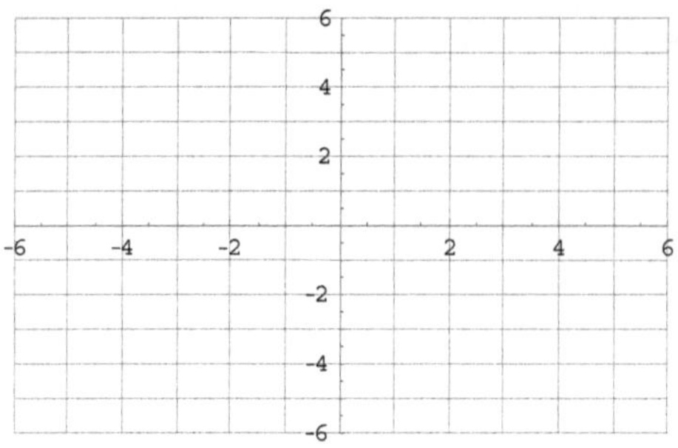

(a) Reflect about the x-axis.

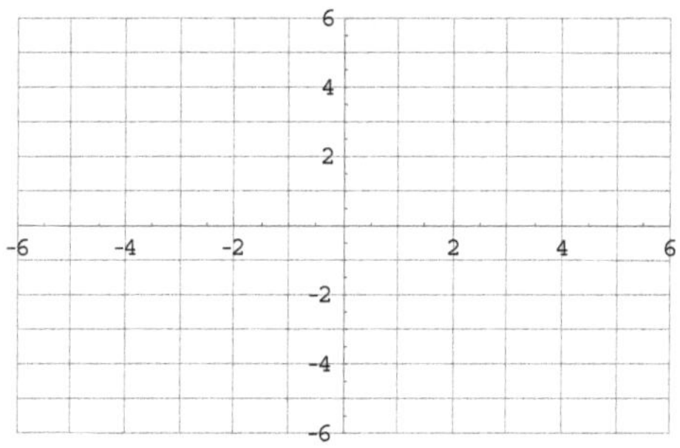

(b) Shift right 3 units.

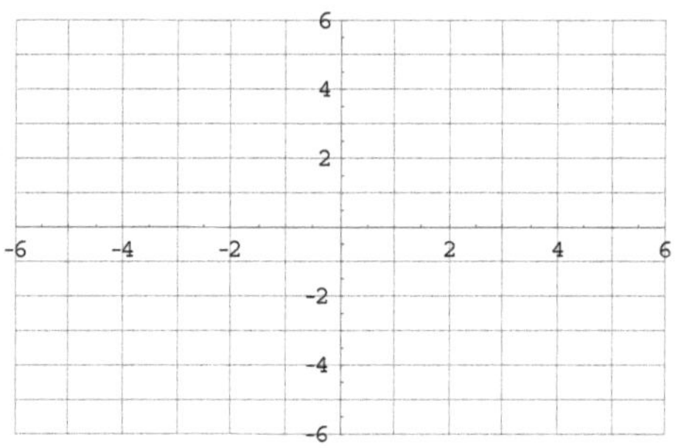

(c) Shift down 2 units.

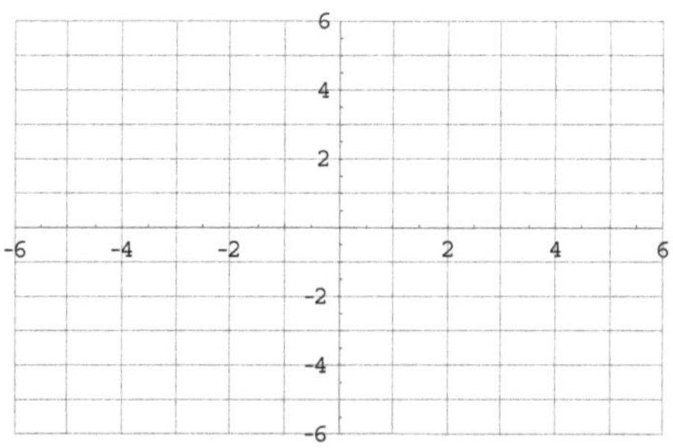

(3) David has available 400 yards of fencing and wishes to enclose a rectangular area.

(a) Express the area A of the rectangle as a function of the width x of the rectangle.

(b) What is the domain of A?

(c) Graph $A = A(x)$ using graphing utility.

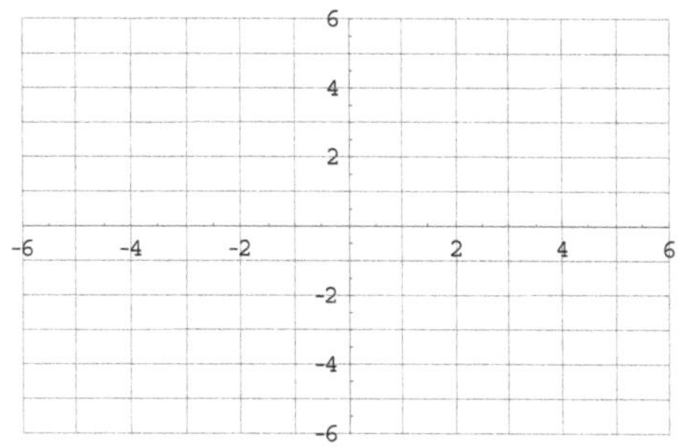

(d) For what value of x is the area largest?

Chapter 12

Quadratic Functions and Models

A quadratic function is a function of the form $f(x) = ax^2 + bx + c$, where a, b, and c are real numbers and $a \neq 0$. The domain of a quadratic function is the set of all real numbers, and its graph is called **parabola**. The parabola is opens upward if $a > 0$ and opens downward if $a < 0$. The vertical line passing through the vertex is called the **axis of symmetry** of the parabola.

If the parabola opens upward, then it has a lowest point and if the parabola opens downward it has a highest point. The lowest and highest points are called the **vertex** of the parabola.

We can find the vertex and axis of symmetry easily if we complete the square of

$$f(x) = ax^2 + bx + c = a(x-h)^2 + k$$

where the point $(h,\ k)$ is the vertex and the vertical line $x = h$ is the equation of the axis of the symmetry.

Therefore, in general for $f(x) = ax^2 + bx + c$, $a \neq 0$,

The vertex: $(-\frac{b}{2a}, f(-\frac{b}{2a}))$

Axis of symmetry: the line $x = -\frac{b}{2a}$

Parabola opens up if $a > 0$; the vertex is a minimum point.

Parabola opens down if $a < 0$; the vertex is a maximum point.

Example: For $f(x) = x^2 + 6x + 5$, find the (a) vertex (b) axis of symmetry (c) intercepts.

Solutions: (a) By completing the square, we have $f(x) = x^2 + 6x + 5 = (x+3)^2 - 4$. So the vertex is $(-3,\ -4)$. (b) The axis of symmetry is $x = -3$. Moreover the parabola opens up and the vertex is the minimum point on the graph. (c) The $y-$intercept is $y = f(0) = 5$, and the $x-$intercepts are:

$0 = (x+3)^2 - 4$
$(x+3)^2 = 4$
$x + 3 = \pm 2$
$x = -3 \pm 2$, i.e., $x = -5$, $x = -1$.

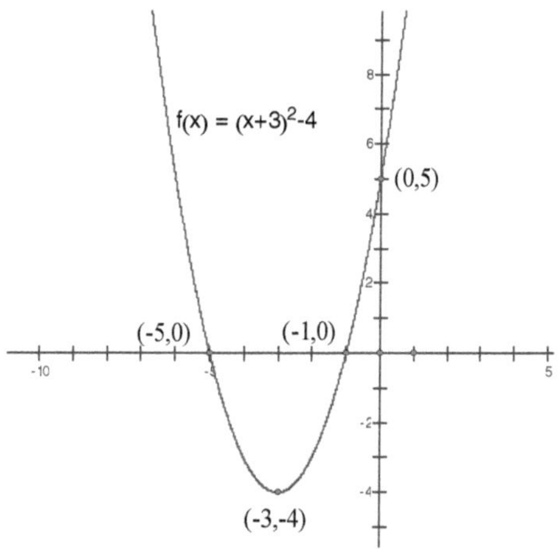

Problem I. A ball is thrown from the top of a building at a velocity of 80 feet per second, its height $h(t)$ in feet is given by the function $h(t) = -16t^2 + 80t + 96$. Answer the following questions:

(1) How tall is the building?
$h(0) = 96$ feet is the height of the building.

(2) What is the maximum height that the ball reaches?
The maximum height is the $y-$coordinate of the vertex.
$h(t) = -16(t - \frac{5}{2})^2 + 196$. So 196 feet is the maximum height.

(3) When does the ball reach its maximum height?
The ball reaches its maximum height at the $x-$coordinate of the vertex. So $t = \frac{5}{2} = 2.5$ seconds.

(4) How high above the building does the ball go?
Max. height-Initial height$= 196 - 96 = 100$ feet.

(5) How long does it take for the ball to reach the height of the building again?
We can use the symmetry of the graph of the parabola to get $t = 5$ seconds, or we can set the function equal to the height of the building 96 ft and solve for t.
Solve: $96 = -16t^2 + 80t + 96$ for t, then $t = 5$ seconds.

(6) When does the ball hit the ground?
When $h(t) = 0$. Solve $0 = -16(t - \frac{5}{2})^2 + 196$, then $t = 6$ or $t = -1$, so the only acceptable answer is $t = 6$ seconds.

Problem II. The price p and the quantity x sold of a certain product obey the demand equation $x = -5p + 100,\ 0 \leq p \leq 20$.

(1) Express the revenue R as a function of x.
$R(x) = \text{price} \times \text{quantity} = \frac{1}{5}(100 - x) \times x = \frac{1}{5}(100x - x^2)$, since $p = \frac{1}{5}(100 - x)$.

(2) What is the revenue if 15 units are sold?
$R(15) = \frac{1}{5}(1500 - 225) = \$255$

(3) What quantity x maximizes revenue? What is the maximum revenue?
It is the $x-$coordinate of the vertex. So $R(x) = -\frac{1}{5}(x - 50)^2 + 500$. So revenue is maximized when $x = 50$ units are sold, and the maximum revenue is \$500.

(4) What price should the company charge to maximize revenue?
At $x = 50$: the price is $p = \dfrac{100 - 50}{5} = 10$. So the company should charge \$10.

Group Exercises 12

Quadratic Functions and Models

(1) Write the quadratic function f in the form $f(x) = a(x-h)^2 + k$ and graph the function.

(a) $f(x) = x^2 - 6x + 10$

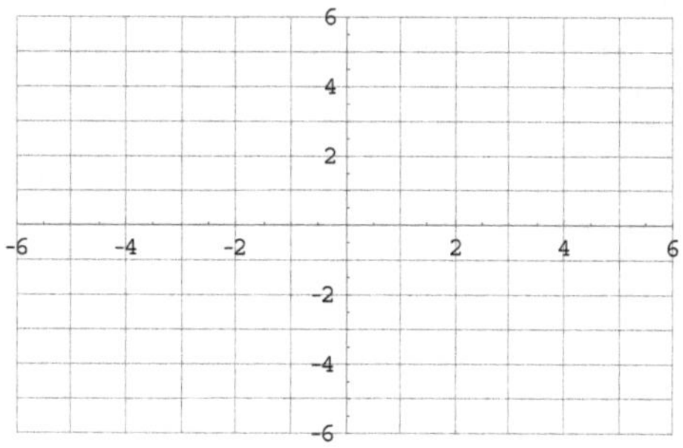

(b) $g(x) = 2x^2 + 6x + 5$

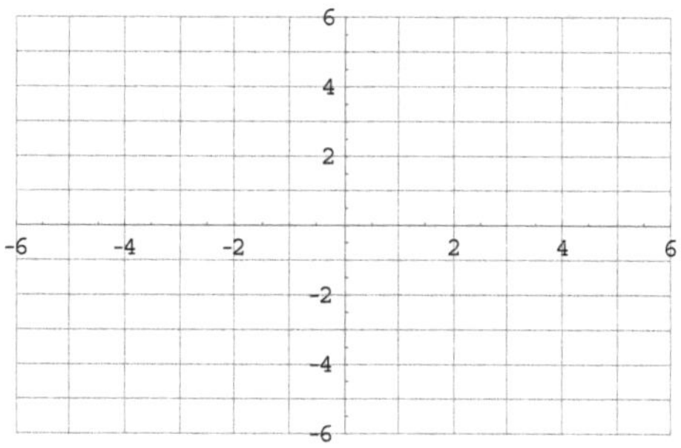

(c) $h(x) = -x^2 - x$

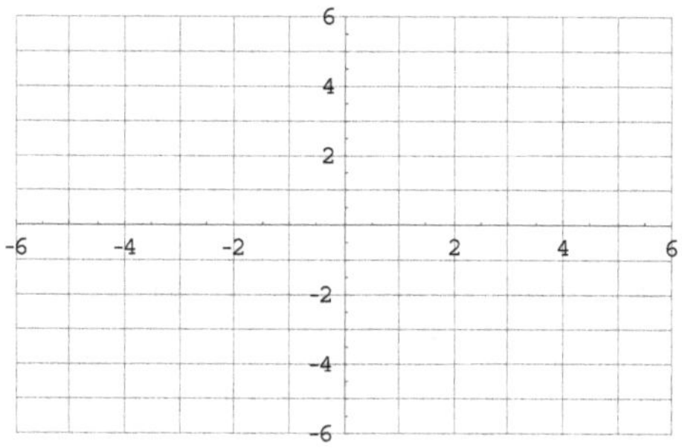

(2) Find the vertex and all intercepts of the parabola. Sketch the graph.

(a) $f(x) = 2x^2 - 4x$

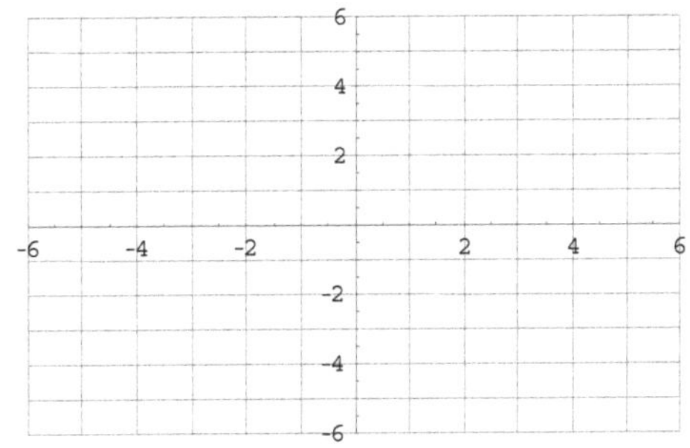

(b) $g(x) = -4x^2 + 4x - 1$

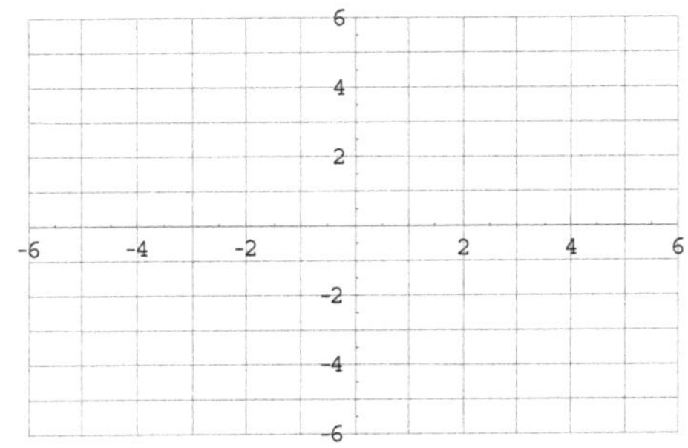

(c) $h(x) = 2x^2 + 12x - 3$

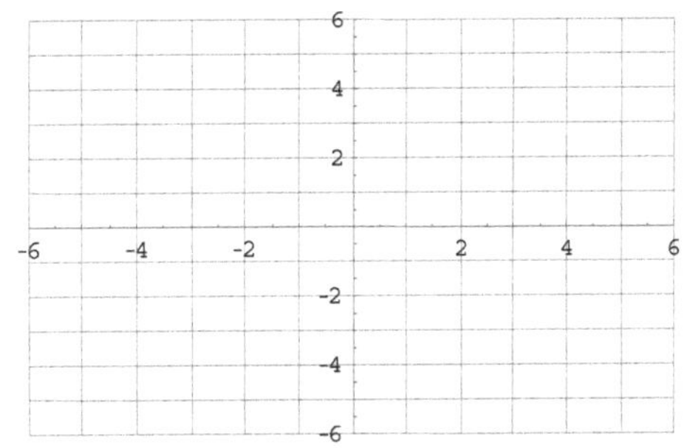

(3) Determine the quadratic function whose graph is given.

(a)

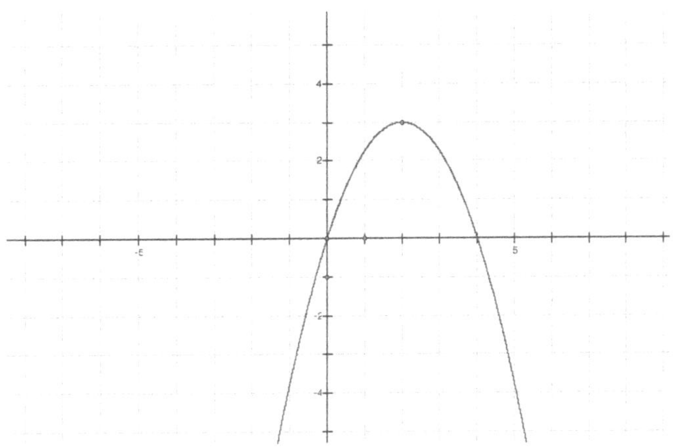

(b)

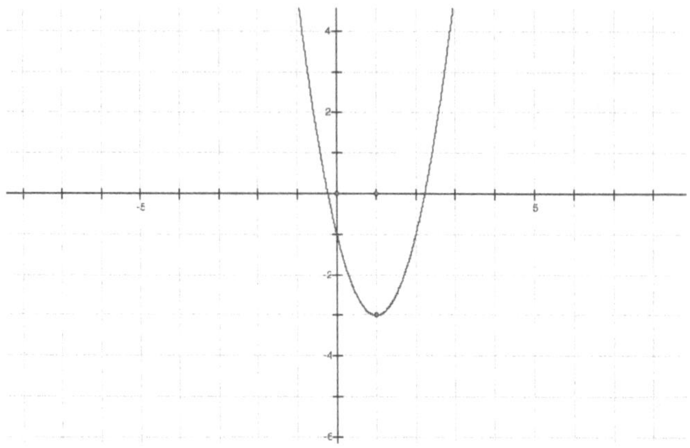

Problem I. Suppose that the manufacturer of a gas cloths dryer has found that, when the unit price is p dollars, the revenue R (in dollars) is $R(p) = -4p^2 + 4000p$. What unit price should be established for the dryer to maximize revenue? What is the maximum revenue?

Problem II. A ball is thrown up from the ground with an initial velocity of 80 ft/s. The height s in feet can be expressed as a function of time t in seconds by $s(t) = 80t - 16t^2$. When does the ball reach its maximum height and what is the maximum height?

Practice Exercises 12

Quadratic Functions and Models

(1) Write the quadratic function f in the form $f(x) = a(x-h)^2 + k$ and graph the function.

(a) $f(x) = -x^2 - 2x - 3$

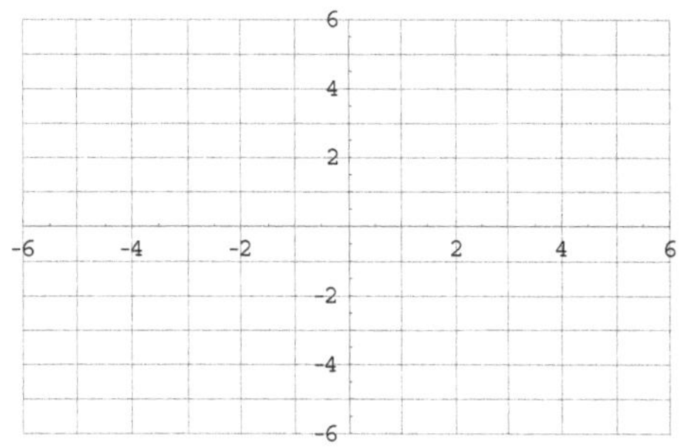

(b) $g(x) = 3x^2 + 12x + 14$

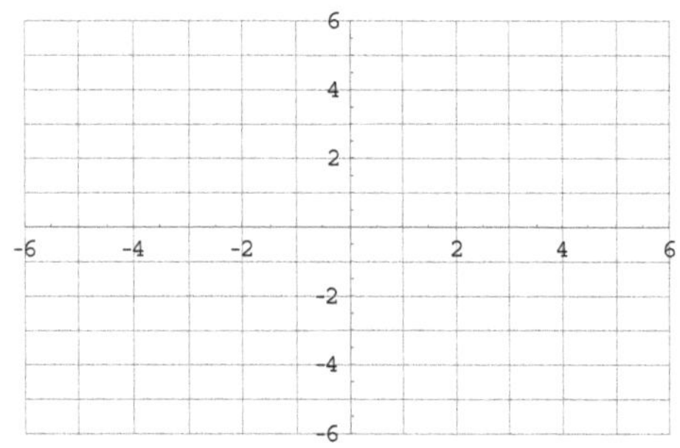

(c) $h(x) = -4x^2 + 4x$

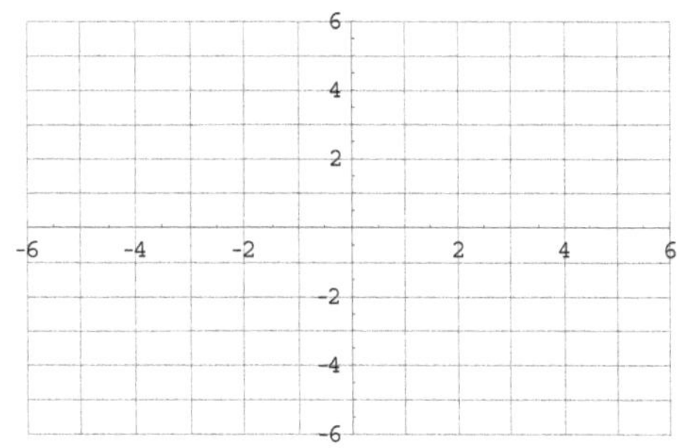

(2) Find the vertex and all intercepts of the parabola. Sketch the graph.

(a) $f(x) = -x^2 - 2x + 3$

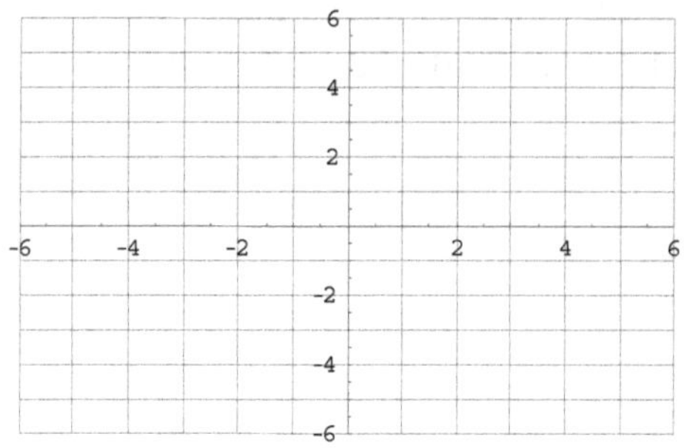

(b) $g(x) = x^2 - x + 1$

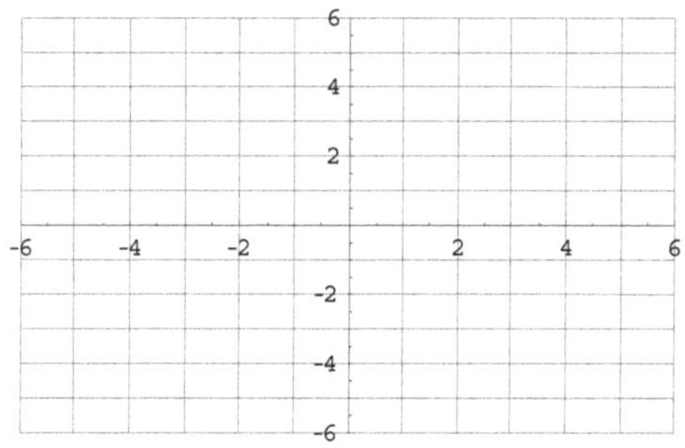

(c) $h(x) = x^2 + 4x + 4$

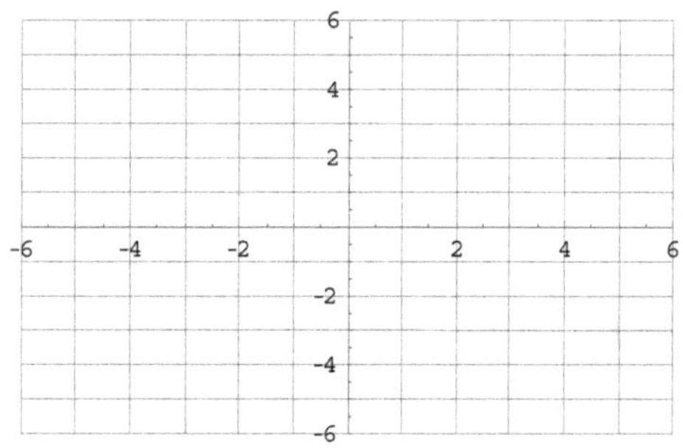

(3) Determine the quadratic function whose graph is given.

(a)

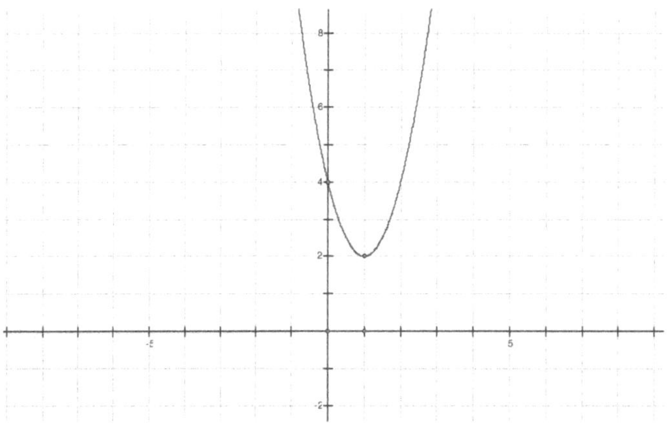

(b)

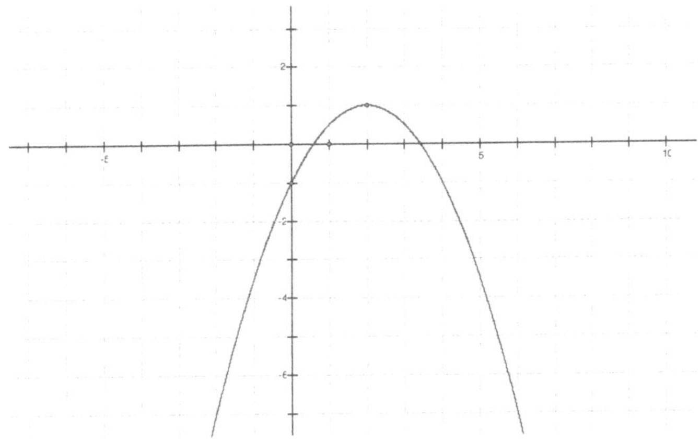

Problem I. Suppose that 320 feet of fencing are available to enclose a rectangular filed and that one side of the field must be given double fencing. What are the dimensions of the rectangle that will yield the largest possible area?

Problem II. The price p and the quantity x sold of a certain product obey the demand equation $x = -20p + 500, \quad 0 \leq p \leq 25$

(a) Express the revenue R as a function of x.

(b) What is the revenue if 20 units are sold?

(c) What quantity x maximizes revenue? What is the maximum revenue?

(d) What price should the company charge to maximize revenue?

Chapter 13

Polynomial Functions

A **polynomial** is a function of the form

$$p(x) = a_n x^n + a_{n-1}x^{n-1} + \cdots + a_1 x + a_0, \ a_n \neq 0$$

where $a_n, \ a_{n-1}, \ \cdots, a_1, \ a_0$ are real numbers and n is a nonnegative integer. The domain is the set of all real numbers. The degree is the largest power of x.

Example 1: The following are polynomial functions in x:
$p(x) = 3x^3 + 2x + 4$ is a polynomial of degree 3,
$q(x) = x^4 - 3x^3 + 2x - 1$, is a polynomial of degree 4
$r(x) = 2x^2 + 3x + 6$, is a polynomial of degree 2
$k(x) = \frac{1}{2}x$ is a polynomial of degree 1, and
$l(x) = 3(x^3 + 2)^4(x^2 + 1)$ is a polynomial of degree 14.

Example 2: The following expressions are not polynomials in x:
$f(x) = 3x^{1/2}$, $g(x) = 1 - \dfrac{2}{x}$, $h(x) = x^{-3} + 3x^2 - x^{1/3} + 5$, $k(x) = \dfrac{x-1}{2x+2}$.

> If f is a polynomial function and r is a real number for which $f(r) = 0$, then r is called a real **zero** of f, or a **root** of f. If r is a (real) zero of f, then
>
> (a) r is an $x-$intercept of the graph of f
>
> (b) $(x - r)$ is a factor of f.

Example 3: Find a polynomial of degree 3 whose zeros are 1, -2, and 3.
Since 1, -2, and 3 are zeros of a polynomial f, then $x - 1$, $x - (-2)$, and $x - 3$ are factors of f. Therefore, a polynomial of degree 3 will be

$$f(x) = a(x - 1)(x + 2)(x - 3)$$

where a is any nonzero real number. The value of a causes a stretch, compression, or reflection, but does not affect the $x-$intercepts. So we can choose a to be 1. Hence

$$f(x) = (x - 1)(x + 2)(x - 3).$$

> If $(x - r)^m$ is a factor of a polynomial function f and $(x - r)^{m+1}$ is not a factor of f, then r is called a **zero of multiplicity** m of f
>
> (a) If r is a zero of even multiplicity, then the graph of f touches the $x-$axis at r.
>
> (b) If r is a zero of odd multiplicity, then the graph of f crosses the $x-$axis at r.

> If f is a polynomial function of degree n, then f has at most $n-1$ turning points.
>
> For a large value of x (either positive or negative), the graph of the polynomial
>
> $f(x) = a_n x^n + a_{n-1}x^{n-1} + \cdots + a_1 x + a_0$ resembles the graph of the power function $y = a_n x^n$.

Example 4: Given the polynomial function $f(x) = 3x^2(x-3)(x+2)^3(x-\frac{1}{2})^4$.

(a) Identify the zeros and their multiplicities.
(b) Find the $x-$ and $y-$intercepts of the graph of f.
(c) Determine whether the graph crosses or touches the $x-$axis at each $x-$intercept.
(d) Determine the degree of f.
(e) Find the number of turning points.
(f) Find the power function that the graph of f resembles for a large values of x.

Solution:

(a), (c) 0 is a zero of multiplicity 2, then the graph touches the $x-$axis at 0.
3 is a zero of multiplicity 1, then the graph crosses the $x-$axis at 3.
-2 is a zero of multiplicity 3, then the graph crosses the $x-$axis at -2.
$\frac{1}{2}$ is a zero of multiplicity 4, then the graph touches the $x-$axis at $\frac{1}{2}$.

(b) $x-$intercepts are the zeros: $0,\ 3,\ -2,\ \frac{1}{2}$
$y-$intercept is $f(0) = 0$.

(d) degree(f)$= n = 10$.

(e) $n - 1 = 9$ is the max. number of turning points.

(f) The power function that the graph of f resembles for a large values of x is $y = 3x^{10}$.

Group Exercises 13

Polynomial Functions

(1) Determine which functions are polynomial functions. For those that are, state the degree. For those that are not, tell why not.

(a) $f(x) = 2 - \dfrac{1}{x}$

(b) $g(x) = 5 + \dfrac{3}{2}x$

(c) $h(x) = 4(x+2)^3(x^2+1)$

(d) $G(x) = \dfrac{x^2 - 9}{x^3}$

(2) Form a polynomial whose zeros and degree are given.

(a) Zeros: -1, 1, 3; degree=3

(b) Zeros: -4, -1, 2, 3; degree=4

(c) Zeros: -1 of multiplicity 1; 3 of multiplicity 2; degree=3

(3) For each polynomial function: (a) List each zero and its multiplicity; (b) Determine whether the graph crosses or touches the x-axis at each x-intercept. (c) Find the power function that the graph of f resembles for large values of $|x|$.

(a) $f(x) = 3(x-7)(x+3)^2$

(b) $g(x) = (x-5)^3(x+4)^2$

(c) $h(x) = -2x^2(x^2-2)$

(4) For each polynomial function, (a) find the x- and y-intercepts of f. (b) determine whether the graph crosses or touches the x-axis at each x-intercept. (c) find the power function that the graph of f resembles for large values of $|x|$. (d) determine the maximum number of the turning points on the graph. (e) use the $x-$intercept(s) to find the intervals on which the graph is above and below the x-axis. (f) plot the points obtained in parts (a)-(e), and use the remaining information to connect them with smooth curve.

(a) $f(x) = x^2(x-3)$

(b) $g(x) = (x-1)(x-2)(x+4)$

(c) $h(x) = x^2(x-2)(x+2)$

Practice Exercises 13

Polynomial Functions

(1) Determine which functions are polynomial functions. For those that are, state the degree. For those that are not, tell why not.

(a) $f(x) = \frac{3}{x} + 1$

(b) $g(x) = -\frac{1}{3}x + 2$

(c) $h(x) = -3(x-1)^2(x^2+1)$

(d) $G(x) = x^{\frac{3}{2}} - \sqrt{x} + \pi x^3$

(2) Form a polynomial whose zeros and degree are given.

(a) Zeros: -3, 0, 4; degree=3

(b) Zeros: -3, -1, 2, 5; degree=4

(c) Zeros: -2 of multiplicity 2; 4 of multiplicity 1; degree=3

(3) For each polynomial function: (a) List each zero and its multiplicity; (b) Determine whether the graph crosses or touches the x-axis at each x-intercept. (c) Find the power function that the graph of f resembles for large values of $|x|$.

(a) $f(x) = 4(x+4)(x+3)^3$

(b) $g(x) = (x - \frac{1}{3})^2(x - 1)^3$

(c) $h(x) = 4x(x^2 - 3)$

(4) For each polynomial function, (a) find the x- and y-intercepts of f. (b) determine whether the graph crosses or touches the x-axis at each x-intercept. (c) find the power function that the graph of f resembles for large values of $|x|$. (d) determine the maximum number of the turning points on the graph. (e) use the $x-$intercept(s) to find the intervals on which the graph is above and below the x-axis. (f) plot the points obtained in parts (a)-(e), and use the remaining information to connect them with smooth curve.

(a) $f(x) = (x + 1)(x + 4)(x - 3)$

(b) $g(x) = 6x^3(x + 4)$

(c) $h(x) = x - x^3$

Chapter 14

Rational Functions and Asymptotes

A **rational function** is a function of the form $R(x) = \dfrac{p(x)}{q(x)}$, where p and q are polynomials and $q \neq 0$. The domain of a rational function is the set of all real numbers except those for which the denominator q is 0. The most basic rational function is $f(x) = \dfrac{1}{x}$, $x \neq 0$. Its domain is the set of all real number except 0, and its graph

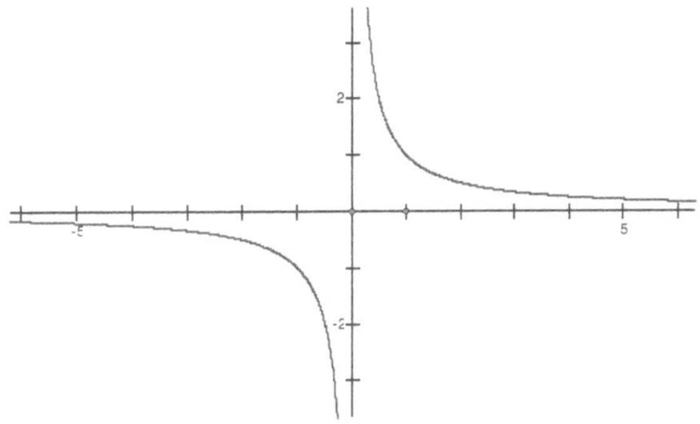

Example 1: Find the domain of each rational function.

(a) $R(x) = \dfrac{1}{x^2 - 9}$: D = All real numbers except 3 and -3, that is $\{x | x \neq 3 \text{ and } -3\}$.

(b) $R(x) = \dfrac{x}{x^2 + 1}$: D = All real numbers, that is $(-\infty,\ \infty)$.

(c) $R(x) = \dfrac{x^2 - 1}{x + 1}$: D $= \{x | x \neq -1\}$

A rational function $R(x) = \dfrac{p(x)}{q(x)}$ is said to be in lowest terms if p and q have no common factors. Then the zeros of the numerator ($p(x) = 0$) are the $x-$intercepts of the graph of R.

Asymptotes

(1) Vertical Asymptotes:

Vertical asymptote, if any, of a rational function $R(x) = \dfrac{p(x)}{q(x)}$, in lowest terms, are the zeros of the denominator $q(x)$. That is, if r is a zero of $q(x)$, then as $x \to r$, $|R(x)| \to 0$ and therefore we say $x = r$ is a vertical asymptote.

Example 2: Find all vertical asymptotes, if any.

(a) $R(x) = \dfrac{x-3}{x+1}$

Since R is in lowest terms and -1 is the zero of the denominator, then the line $x = -1$ is the vertical asymptote. (See the graph below).

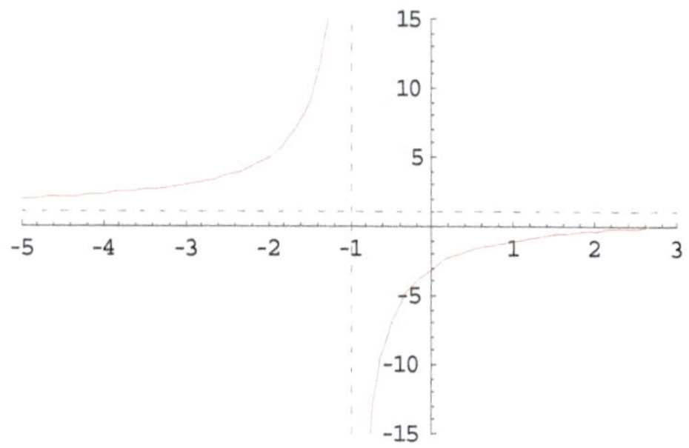

(b) $H(x) = \dfrac{x^2-16}{x^2-11x+28}$

$$H(x) = \frac{x^2-16}{x^2-11x+28} = \frac{(x-4)(x+4)}{(x-4)(x-7)} = \frac{x+4}{x-7},\ x \neq 4.$$

Hence, 7 is the only zero of the denominator and therefore, the line $x = 7$ is the only vertical asymptote.

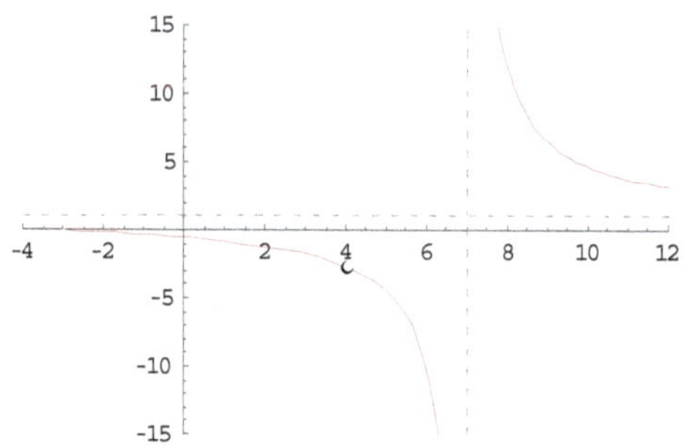

(c) $F(x) = \dfrac{x}{x^2+4}$

F is in lowest terms and the denominator has no real zeros, therefore the graph of F has no vertical asymptotes.

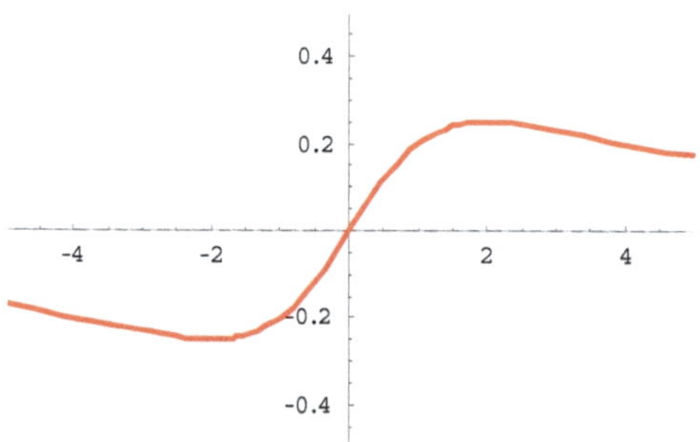

(d) $G(x) = \dfrac{x^2}{x^2 - 25}$

G is in lowest terms and the zeros of the denominator $x^2 - 25$ are -5 and 5. Hence, the lines $x = 5$ and $x = -5$ are the vertical asymptotes.

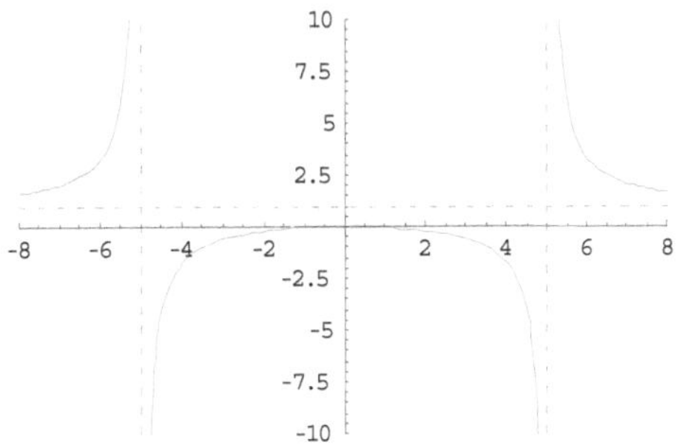

(2) Horizontal Asymptotes:

(a) If the rational function is proper, that is, if the degree of the numerator is less than the degree of the denominator, then the line $y = 0$ is a horizontal asymptote. (As $x \to \pm\infty$ the value of $R(x)$ approaches 0.)

Example 3: Find the horizontal asymptote, if any, of the graph of

$$R(x) = \frac{x - 5}{x^2 - 3x - 10}$$

The rational function R is proper (the degree of the numerator is, 1, is less than the degree of the denominator, 2). Therefore, the line $y = 0$ is a horizontal asymptote of the graph of R.

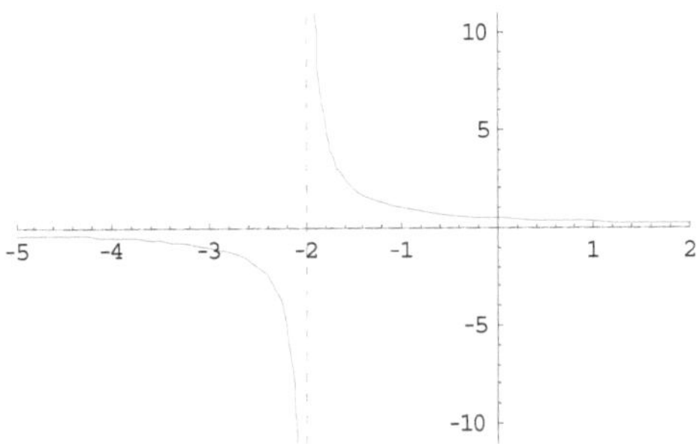

(b) If the rational function $\dfrac{p(x)}{q(x)}$ is improper, that is, if the degree of the numerator is greater than or equal to the degree of the denominator, then we must use long division to write the rational function as

$$\frac{p(x)}{q(x)} = f(x) + \frac{r(x)}{q(x)}$$

where $f(x)$ is a polynomial and $\frac{r(x)}{q(x)}$ is proper rational function and hence $\frac{r(x)}{q(x)} \to 0$. Therefore, we have the following three possibilities:

(1) If $f(x) = b$, then the line $y = b$ is a horizontal asymptotes of the graph of R.

(2) If $f(x) = ax + b$, $a \neq 0$, then the line $y = ax + b$ is an oblique asymptote of the graph of R.

(3) In all other cases, the graph of f will approach the graph of R and therefore there are no horizontal or oblique asymptotes.

Example 4: Find the horizontal or oblique asymptotes, if any, of the graph of

$$R(x) = \frac{x^2}{x^2 - 1}$$

Solution: R is improper and by the long division $R(x) = \frac{x^2}{x^2-1} = 1 + \frac{1}{x^2-1}$. Then as $x \to \infty$ or $x \to -\infty$, $\frac{1}{x^2-1} \to 0$. So, $R(x) \to 1$, and we conclude that $y = 1$ is a horizontal asymptote. (See the graph below).

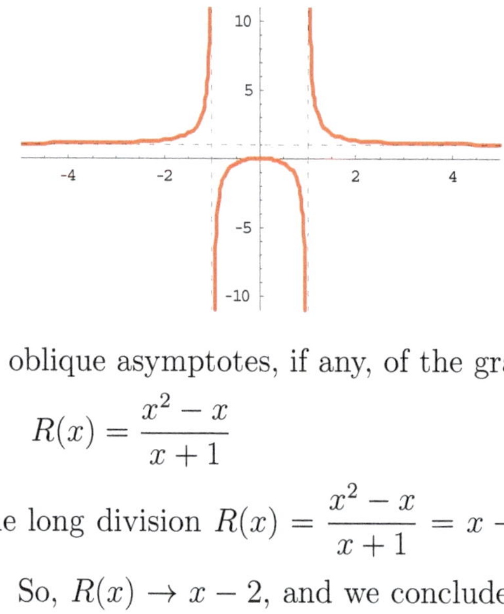

Example 5: Find the horizontal or oblique asymptotes, if any, of the graph of

$$R(x) = \frac{x^2 - x}{x + 1}$$

Solution: R is improper and by the long division $R(x) = \frac{x^2 - x}{x+1} = x - 2 + \frac{2}{x+1}$. Then as $x \to \infty$ or $x \to -\infty$, $\frac{2}{x+1} \to 0$. So, $R(x) \to x - 2$, and we conclude that $y = x - 2$ is oblique asymptote.

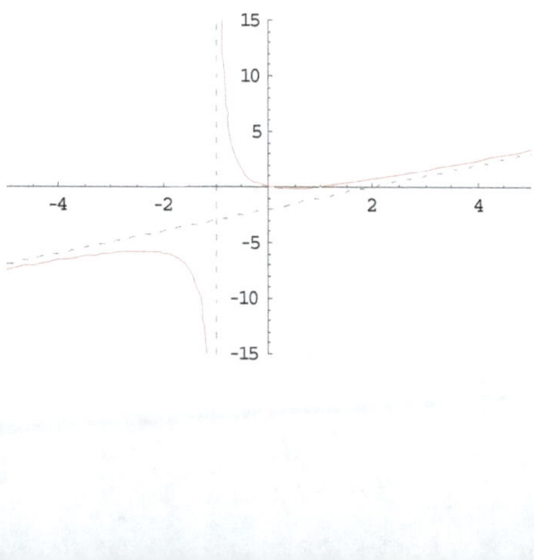

Example 6: Find the horizontal or oblique asymptotes, if any, of the graph of

$$H(x) = \frac{2x^5 - x^3 + 2}{x^3 - 1}$$

Solution: H is improper and by the long division $H(x) = \dfrac{2x^5 - x^3 + 2}{x^3 - 1} = 2x^2 - 1 + \dfrac{2x^2 + 1}{x^3 - 1}$.
Then as $x \to \infty$ or $x \to -\infty$, $\dfrac{2x^2 + 1}{x^3 - 1} \to 0$. So, $H(x) \to 2x^2 - 1$ as $x \to \pm\infty$. That is the graph of H will look like the graph of $y = 2x^2 - 1$. Since $y = 2x^2 - 1$ is not a linear function, H has no horizontal or oblique asymptotes.

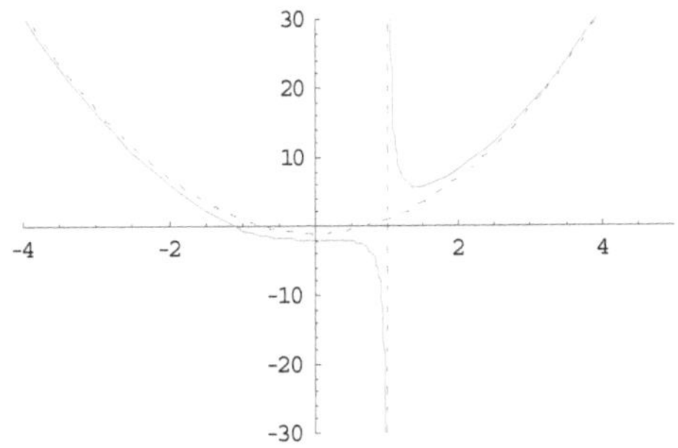

Group Exercises 14

Rational Functions and Asymptotes

(1) Find the domain of each rational function:

(a) $R(x) = \dfrac{4x}{x-3}$

(b) $F(x) = \dfrac{3x(x-1)}{2x^2-5x-3}$

(c) $H(x) = \dfrac{3x^2+x}{x^2+4}$

(2) Find the vertical, horizontal, and oblique asymptotes, if any, of each rational function.

(a) $R(x) = \dfrac{3x}{x+4}$

(b) $T(x) = \dfrac{x^3}{x^4-1}$

(c) $H(x) = \dfrac{3x^4 + 4}{x^3 + 3x}$

(d) $G(x) = \dfrac{x^3 + 4x^2 + 3x}{x^2 - 25}$

(3) Graph each rational function using transformations.

(a) $F(x) = \dfrac{2}{(x-3)^2} + 1$

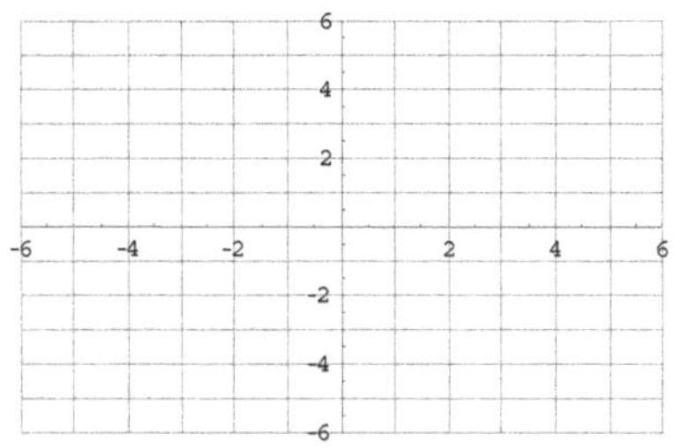

(b) $F(x) = 2 - \dfrac{1}{x+1}$

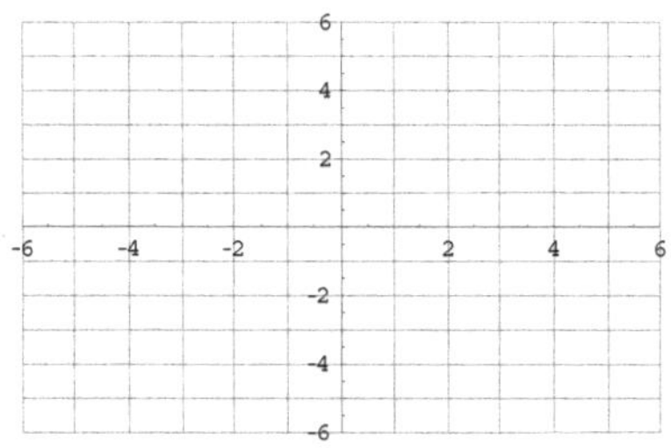

Problem I. The concentration C of a certain drug in a patient's bloodstream t minutes after injection is given by

$$C(t) = \frac{50t}{t^2 + 25}$$

(a) Find the horizontal asymptote of $C(t)$. What happens to the concentration of the drug as t increases?

(b) Graph $C(t)$

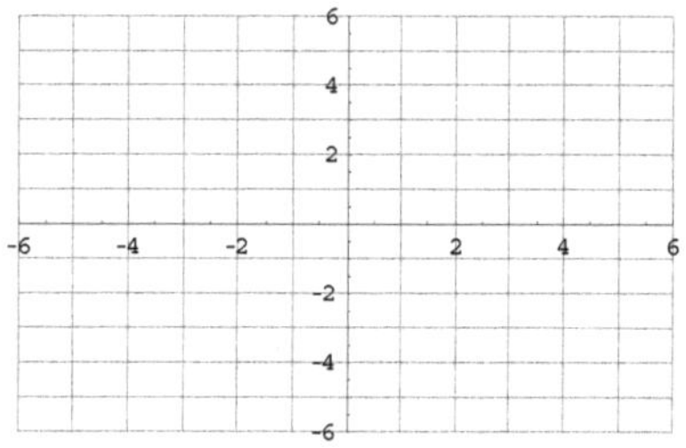

(c) Determine the time at which the concentration is highest.

Practice Exercises 14

Rational Functions and Asymptotes

(1) Find the domain of each rational function:

(a) $R(x) = \dfrac{5x^2}{3+x}$

(b) $G(x) = \dfrac{-4x^2}{(x-2)(x+4)}$

(c) $H(x) = \dfrac{x^2-4}{x^2+4x+4}$

(2) Find the vertical, horizontal, and oblique asymptotes, if any, of each rational function.

(a) $R(x) = \dfrac{2x^2+1}{x^2-4}$

(b) $T(x) = \dfrac{x^2}{3x-3}$

(c) $H(x) = \dfrac{x-1}{2x^3 - 2x}$

(d) $G(x) = \dfrac{x^4 + 2x^2 + 1}{x^2 - x + 1}$

(3) Graph each rational function using transformations.

(a) $F(x) = \dfrac{1}{x-1} + 2$

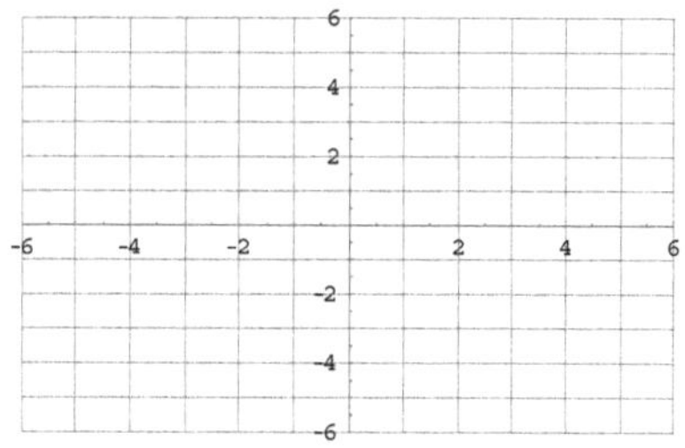

(b) $F(x) = \dfrac{-1}{x^2 + 4x + 4}$

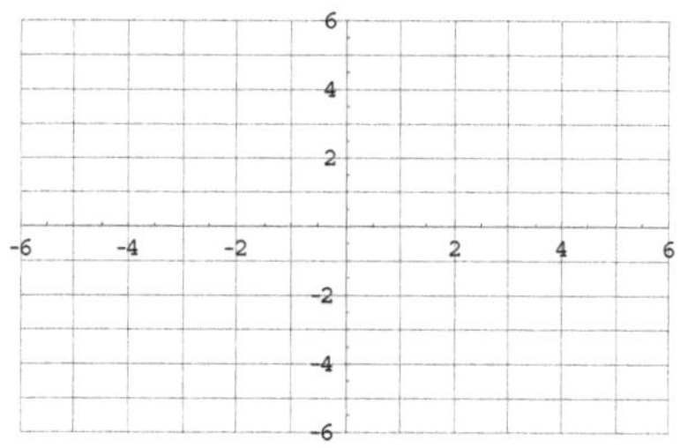

Problem I. The concentration C of a certain drug in a patient's bloodstream t minutes after injection is given by

$$C(t) = \frac{t}{2t^2 + 1}$$

(a) Find the horizontal asymptote of $C(t)$. What happens to the concentration of the drug as t increases?

(b) Graph $C(t)$

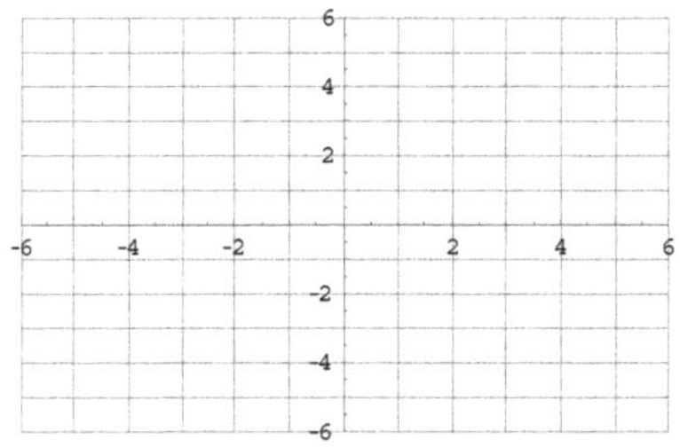

(c) Determine the time at which the concentration is highest.

Chapter 15

Exponential Functions and Equations

An exponential function is a function of the form $f(x) = Ca^x$, where $a > 0$, $a \neq 1$, $C \neq 0$, and x is any real number. The domain is the set of all real numbers. The range is the set of all positive real numbers. An exponential function has a horizontal asymptote at $y = 0$. A logarithmic function is the inverse of an exponential function. That is,

$$\boxed{y = a^x \quad \text{if and only if} \quad x = \log_a y.}$$

Natural Exponential Function: The function $f(x) = e^x$ is called the natural exponential function, where $e \approx 2.718$. Its inverse is the natural logarithm:

$$\boxed{y = e^x \quad \text{if and only if} \quad x = \ln y.}$$

Example 1: The following are exponential functions:

$$f(x) = 2^x, \quad g(x) = \left(\tfrac{1}{2}\right)^x, \quad h(x) = e^x.$$

Example 2: The following are not exponential functions:

$$x^2, \quad (-2)^x, \quad 2^{x^2}, \quad x^x.$$

Laws of Exponents: If m, n, a, b are real numbers with $a > 0$ and $b > 0$, then

(1) $a^m \cdot a^n = a^{m+n}$ (2) $\dfrac{a^m}{a^n} = a^{m-n}$ (3) $(a^m)^n = a^{mn}$

(4) $a^{-n} = \dfrac{1}{a^n} = \left(\dfrac{1}{a}\right)^n$ (5) $(ab)^m = a^m b^m$ (6) $a^0 = 1$ (7) $1^m = 1$

Example 3: If $f(x) = a^x$, then:

(a) There is no x-intercept, the y-intercept is $(0, 1)$.
(b) The domain is the set of all real numbers, or $(-\infty, \infty)$, the range is the of postive real numbers, or $(0, \infty)$.
(c) If $a > 1$, the function is increasing, and if $0 < a < 1$, the function is decreasing.
(d) The $x-$axis $(y = 0)$ is a horizontal asymptote as $x \to -\infty$.
(e) The graph of f contains the points $(-1, \dfrac{1}{a})$, $(0, 1)$, and $(1, a)$.

Example 4: Graph $y = 2^x$ and $y = 2^{-x} = (\frac{1}{2})^x$.

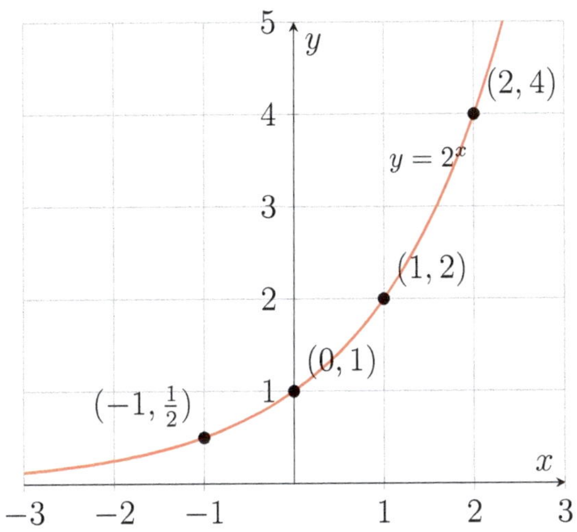

Increasing, asymptote $y = 0$

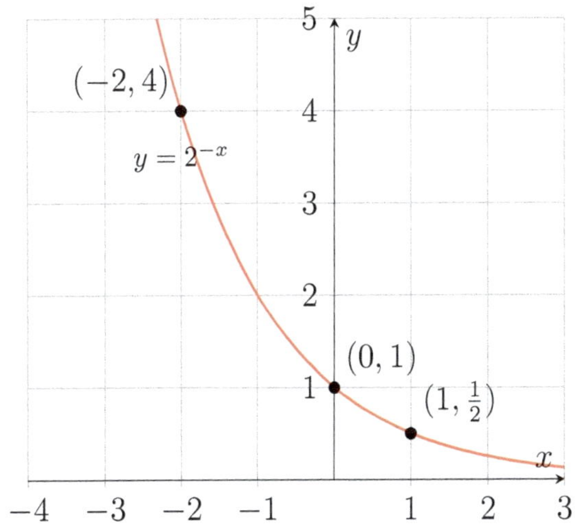

Decreasing, asymptote $y = 0$

Example 5: Graph $y = e^x$: and $y = e^{-x}$.

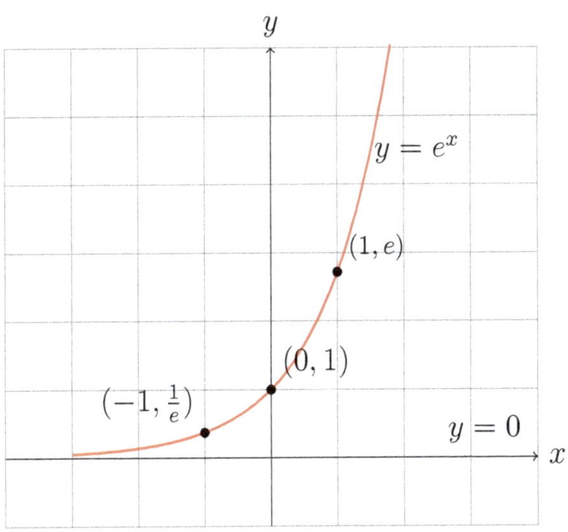

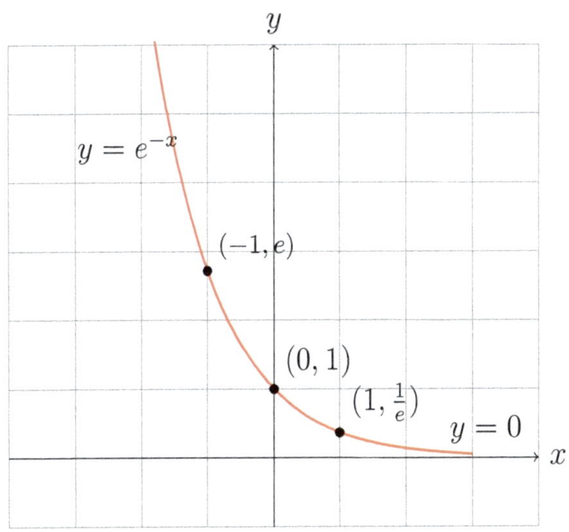

The graph of $y = e^x$ has the same general shape as $y = a^x$ with $a > 1$, and the graph of $y = e^{-x}$ has also similar shape as $y = a^{-x}$ with $0 < a < 1$ which is decreasing. There is no vertical asymptotes for those functions, but they do have a horizontal asymptote $y = 0$. The graph of $y = e^x$ increases as $x \to \infty$, and the graph of $y = e^{-x}$ decreases and approches 0 as

$x \to \infty$. Both functions pass through the point $(0,\ 1)$. Finally, both graphs are continuous, with no corners or gaps.

Solving Exponential Equations

$$\boxed{\text{If } a^x = a^y, \quad \text{then} \quad x = y.}$$

Example 6: Solve $2^x = 8$.

Solution: Since $8 = 2^3$, then $2^x = 2^3 \Rightarrow x = 3$.

Example 7: Solve $3^{x+1} = 27$.

Solution: Since $27 = 3^3$, then $3^{x+1} = 3^3 \Rightarrow x + 1 = 3$, and thus $x = 2$.

Example 8: Solve $8^{-x+14} = 16^x$.

Solution: Since $8 = 2^3$, and $16 = 2^4$, then $8^{-x+14} = (2^3)^{(-x+14)} = 2^{(-3x+42)}$ and $16^x = (2^4)^x = 2^{4x} \Rightarrow 2^{(-3x+42)} = 2^{4x} \Rightarrow -3x + 42 = 4x$, and hence $x = 6$.

Example 9: Graph $f(x) = 2^{x-3} + 1$ using transformations, and determine the domain, range, and horizontal asymptote of f.

Solution: Begin with the graph of $y = 2^x$, then shift right 3 units ($y = 2^{x-3}$), finally shift up 1 unit ($y = 2^{x-3} + 1$).

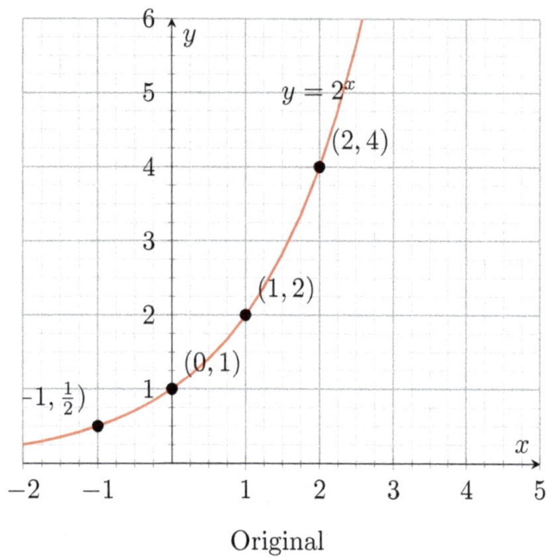

Original

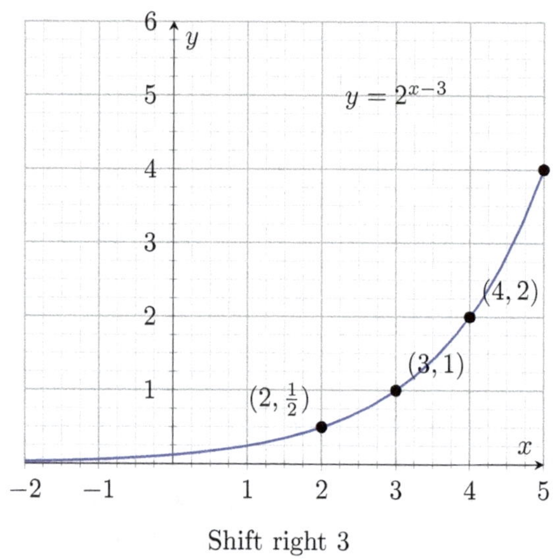

Shift right 3

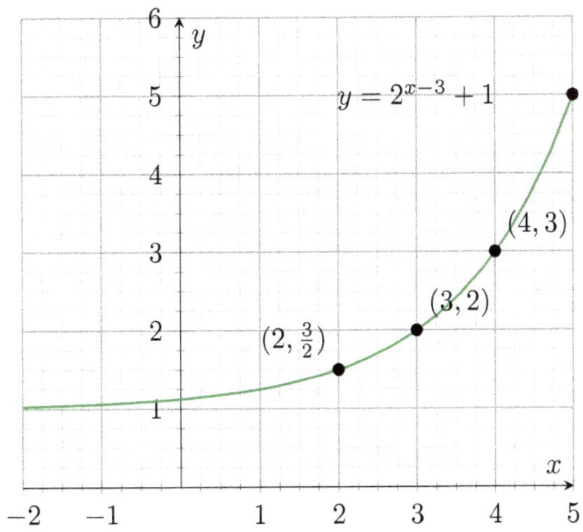

Right 3, Up 1

Group Exercises 15

Exponential Functions and Equations

(1) Determine which functions are exponential functions. For those that are, state the base. For those that are not, tell why not.

(a) $f(x) = 2^x$

(b) $g(x) = x^2$

(c) $h(x) = \left(\frac{1}{3}\right)^x$

(d) $G(x) = (-2)^x$

(e) $k(x) = e^x$

(2) Simplify each expression.

(a) $2^3 \cdot 2^4$

(b) $\frac{5^6}{5^2}$

(c) $(3^2)^3$

(d) $e^2 \cdot e^3$

(e) $\frac{e^5}{e^2}$

(3) Solve each exponential equation.

(a) $3^x = 81$

(b) $2^{x+2} = 32$

(c) $9^{-x+15} = 27^x$

(d) $5^{x^2+8} = 125^{2x}$

(e) $e^{3x} = e^{2-x}$

(f) $4^x \cdot 2^{x^2} = 16^2$

(4) Use transformations to graph each function. Determine the domain, range, and horizontal asymptote of each function.

(a) $f(x) = 3^x - 2$

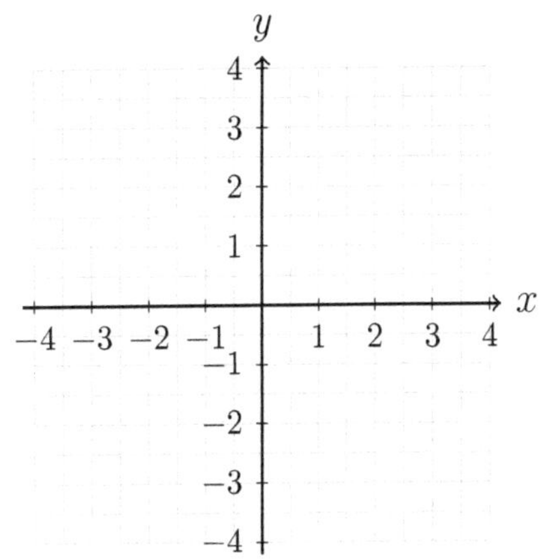

(b) $f(x) = -3^x + 2$

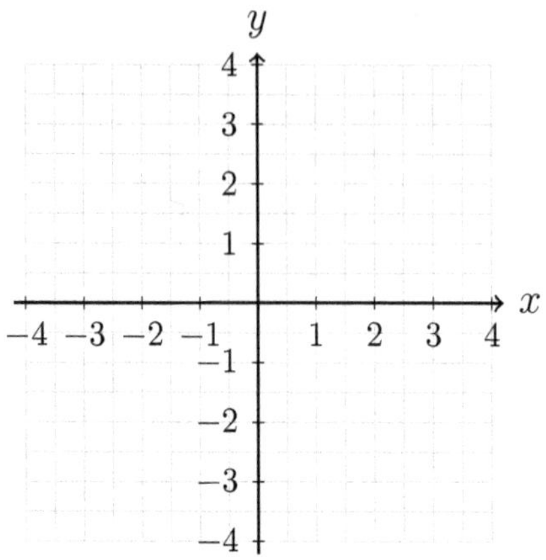

(c) $f(x) = 1 - 2^{x+3}$

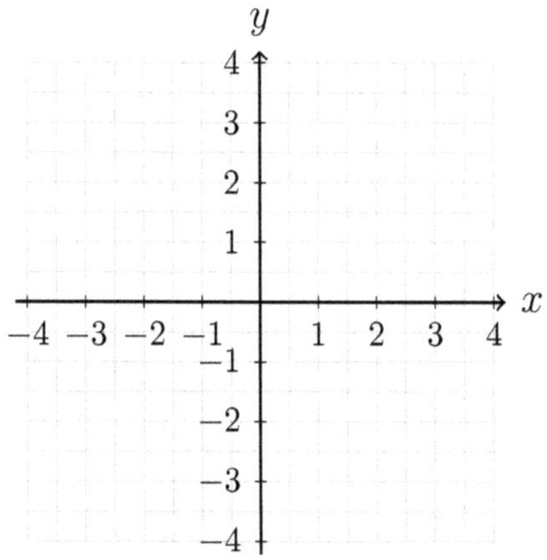

(d) $f(x) = 5 - e^{-x}$

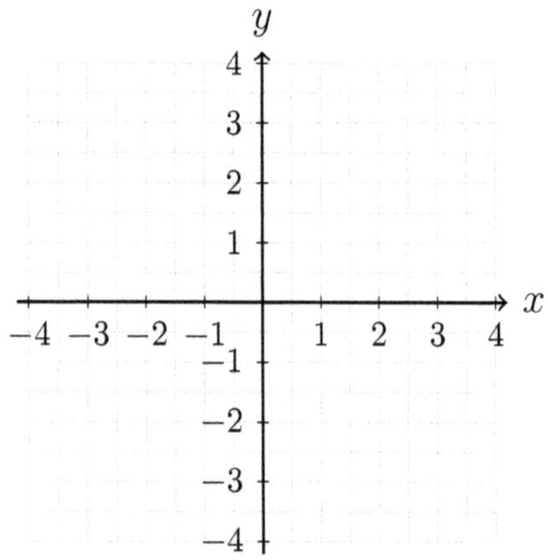

Practice Exercise 15

Exponential Functions and Equations

(1) Solve each exponential equation.

(a) $2^{-x} = 16$

(b) $3^{x^2-7} = 27^{2x}$

(c) $e^{x^2} = e^{3x} \cdot \frac{1}{e^2}$

(d) $(\frac{1}{4})^x = \frac{1}{64}$

(e) $e^{3x} = e^{2-x}$

(f) $4^x \cdot 2^{x^2} = 16^2$

(2) Use transformations to graph each function. Determine the domain, range, and horizontal asymptote of each function.

(a) $f(x) = 3^{x-2}$

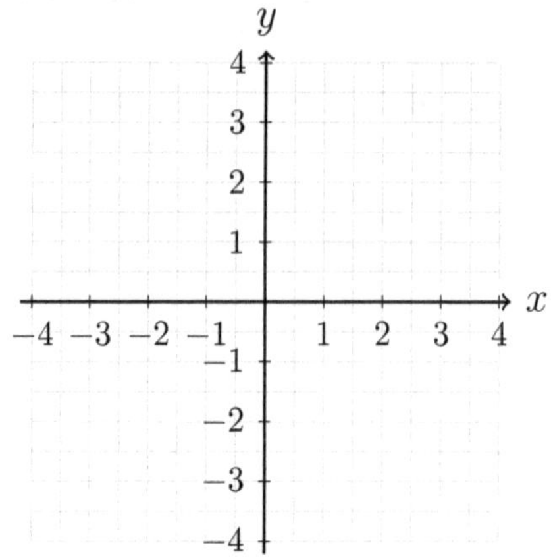

(b) $f(x) = 3^{-x} + 2$

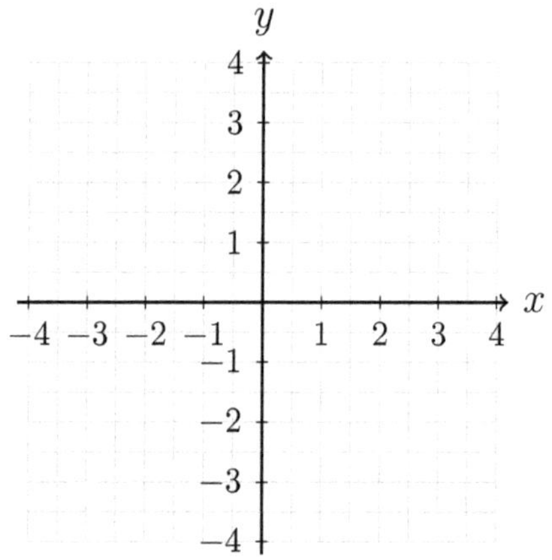

(c) $f(x) = 4 - 2^{x+2}$

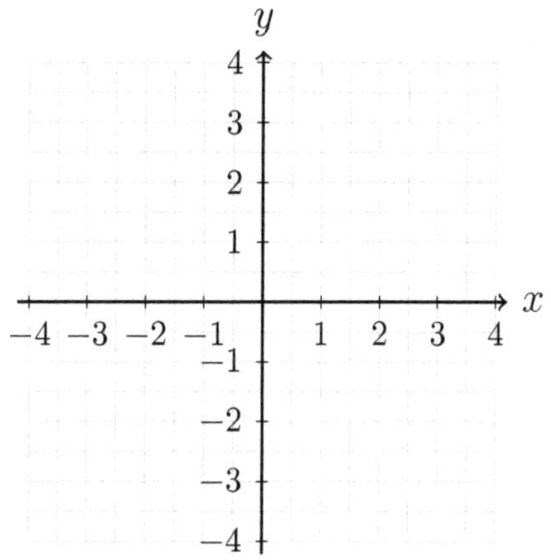

(d) $f(x) = e^{x+2} + 3$

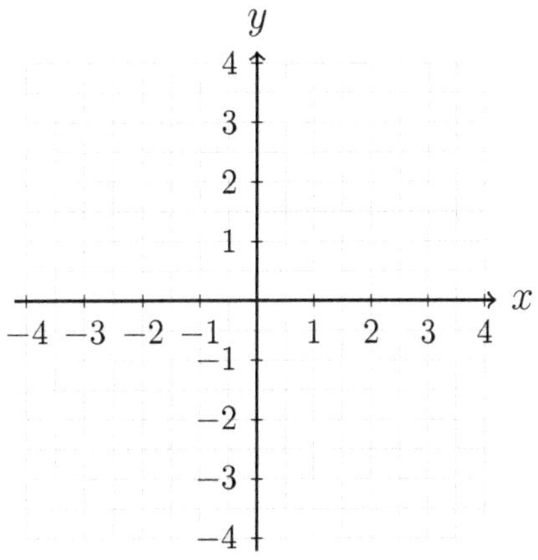

(3) Suppose that $g(x) = 4^x + 2$.

(a) What is $g(-1)$? What point is on the graph of g?

(b) If $g(x) = 66$, what is x? What point is on the graph of g?

Chapter 16

Logarithmic Functions and Equations

A logarithmic function is a function of the form

$$f(x) = \log_a x$$

where $a > 0$, $a \neq 1$, and $x > 0$. The domain is the set of all positive real numbers. The range is the set of all real numbers. A logarithmic function is the inverse of an exponential function. That is,

$$\boxed{y = \log_a x \quad \text{if and only if} \quad x = a^y.}$$

Example 1: Change each exponential statement to an equivalent statement involving a logarithm.

(a) $a^3 = 27$ (b) $e^a = 5$ (c) $3^{-2} = c$.

Solution:

(a) If $a^3 = 27$, then $3 = \log_a 27$.

(b) If $e^a = 5$, then $a = \log_e 5 = \ln 5$.

(c) If $3^{-2} = c$, then $-2 = \log_3 c$.

Example 2: Change each logarithmic statement to an equivalent statement involving an exponent.

(a) $\log_a 8 = 3$ (b) $\log_e b = -1$ (c) $\log_2 16 = c$.

Solution:

(a) If $\log_a 8 = 3$, then $a^3 = 8$.

(b) If $\log_e b = -1$, then $e^{-1} = b$.

(c) If $\log_2 16 = c$, then $2^c = 16$.

Example 3: Find the exact value.

(a) $\log_2 32$ (b) $\log_3 \dfrac{1}{81}$.

Solution:

(a) Let $y = \log_2 32$, then $2^y = 32 = 2^5 \Rightarrow y = 5$.

(b) Let $y = \log_3 \frac{1}{81}$, then $3^y = \frac{1}{81} = \frac{1}{3^4} = 3^{-4} \Rightarrow y = -4$.

Example 4: Find the domain of the logarithmic functions.

(a) $f(x) = \log_2(x-1)$ (b) $g(x) = \log_{10}\left(\dfrac{x+3}{x-3}\right)$.

Solution:

(a) The domain of f consists of all x for which $x - 1 > 0$, implies that $x > 1$. In interval notation, the domain of f is $(1, \infty)$.

(b) The domain of g consists of all x for which $\dfrac{x+3}{x-3} > 0$, that is, when both $x+3$ and $x-3$ positive or when both negative.

$$x + 3 = 0 \Rightarrow x = -3, \qquad x - 3 = 0 \Rightarrow x = 3.$$

These points divide the real line into intervals:

$$(-\infty, -3), \quad (-3, 3), \quad (3, \infty)$$

Now, test each interval:

$$x = -4: \ \frac{-4+3}{-4-3} = \frac{-1}{-7} > 0$$
$$x = 0: \ \frac{0+3}{0-3} = \frac{3}{-3} < 0$$
$$x = 4: \ \frac{4+3}{4-3} = \frac{7}{1} > 0.$$

Therefore, the solution is

$$x < -3 \quad \text{or} \quad x > 3.$$

In interval notation:

$$\boxed{(-\infty, -3) \cup (3, \infty)}$$

If the base of a logarithmic function is the number e, where $e \approx 2.718281828\ldots$, then $\log_e x = \ln x$ is the **natural logarithm function.**

$$\boxed{y = \ln x \quad \text{if and only if} \quad x = e^y}$$

The logarithmic functions $y = \ln(x)$ and the exponential function $y = e^x$ are inverses of each other, so the graph of $y = \ln(x)$ is a reflection of the graph of $y = e^x$ about the line $y = x$.

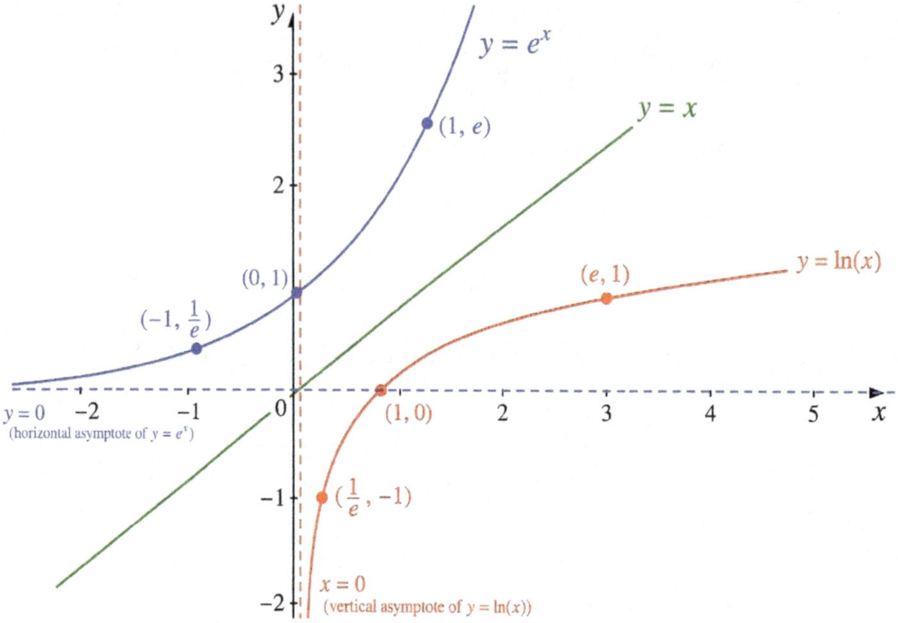

Properties of Logarithms:

We assume M, N, a, and r are real numbers, $a \neq 1$. $b \neq 1$, $M > 0$, and $N > 0$.

$$\text{If } \log_a M = \log_a N, \quad \text{then} \quad M = N$$
$$\text{If } M = N, \quad \text{then} \quad \log_a M = \log_a N.$$

$$\log_a(MN) = \log_a M + \log_a N$$
$$\log_a\left(\tfrac{M}{N}\right) = \log_a M - \log_a N$$
$$\log_a(M^r) = r \log_a M$$
$$\log_a M = \frac{\log_b M}{\log_b a} \quad \text{(Change-of-base formula)}$$
$$\log_a 1 = 0$$
$$\log_a a = 1$$

Solving Logarithmic Equations:

We solve logarithmic equations by changing a logarithmic expression to an exponential expression.

$$\boxed{y = \log_a x \quad \text{is equivalent to} \qquad x = a^y, \quad a > 0, \quad a \neq 1}$$

Example 5: Solve: $\log_2(x+6) = 4$.

Solution: Write the logarithmic equation as an equivalent exponential equation:

$$\log_2(x+6) = 4 \Rightarrow x + 6 = 2^4 = 16 \Rightarrow x = 10.$$

Example 6: Solve: $-2\log_4 x = \log_4 9$.

Solution:

$$\begin{aligned} -2\log_4 x &= \log_4 9 \\ \log_4 x^{-2} &= \log_4 9 \\ x^{-2} &= 9 \\ x^2 &= \frac{1}{9} \\ x &= \pm\frac{1}{3}. \end{aligned}$$

Since the domain of variable is $x > 0$, which implies that $\frac{-1}{3}$ must be discarded. We can now check whether $\frac{1}{3}$ is a solution:

$$\begin{aligned} -2\log_4 \tfrac{1}{3} &\stackrel{?}{=} \log_4 9 \\ \log_4(\tfrac{1}{3})^{-2} &\stackrel{?}{=} \log_4 9 \\ \log_4(3)^2 &\stackrel{?}{=} \log_4 9 \\ \log_4 9 &= \log_4 9. \end{aligned}$$

Hence, the solution $\{\frac{1}{3}\}$.

Example 7: Solve: $\ln(x+1) - \ln x = 2$.

Solution:

$$\begin{aligned} \ln(x+1) - \ln x &= 2 \\ \ln(\frac{x+1}{x}) &= 2 \end{aligned}$$

$$\frac{x+1}{x} = e^2$$
$$1 + x^{-1} = e^2$$
$$x^{-1} = e^2 - 1$$
$$x = \frac{1}{e^2-1} > 0.$$

Since the domain consists of $x > 0$, then $\frac{1}{e^2-1} > 0$ is a solution.

Example 8: Use logarithms to solve the exponential equation $e^{3x} = 10$.

Solution: Take the ln of both sides:

$$e^{3x} = 10$$
$$\ln e^{3x} = \ln 10$$
$$\text{but } \ln e^{3x} = 3x$$
$$\text{since } \ln e = 1, \text{ then}$$
$$3x = \ln 10$$
$$x = \frac{\ln 10}{3}.$$

The solution set is $\{\frac{\ln 10}{3}\}$.

In general, if $f(x) = \log_a x$, then:

(a) The graph has a vertical asymptote at $x = 0$.
(b) The x-intercept is $(1, 0)$.
(c) There is no y-intercept.
(d) If $a > 1$, the function is increasing.
(e) If $0 < a < 1$, the function is decreasing.

Example 9 : Graph of $y = \log_2 x$:

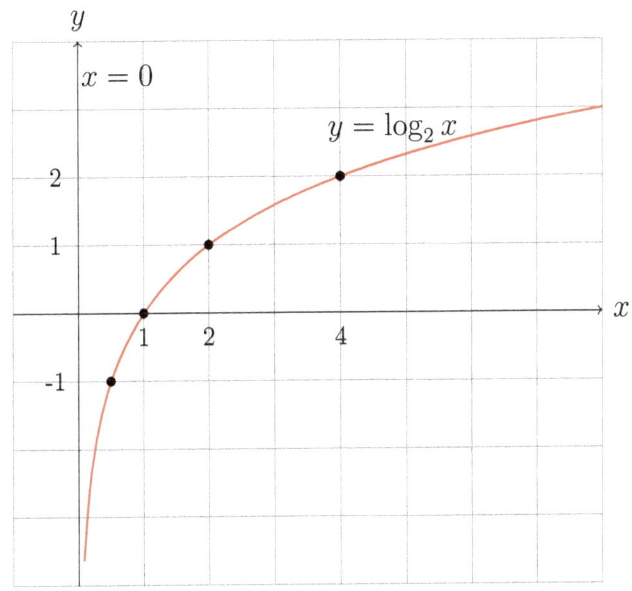

The graph of $y = \log_2 x$ is increasing, has domain $(0, \infty)$, range $(-\infty, \infty)$, and a vertical asymptote at $x = 0$.

Example 10: Graph of $y = \ln x$:

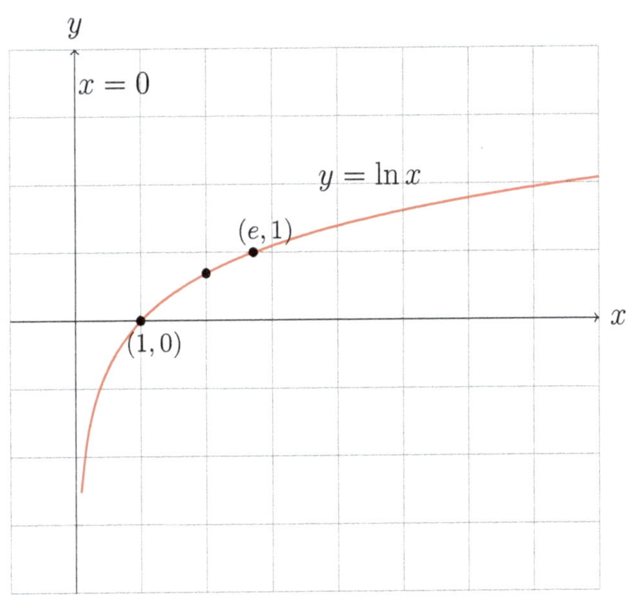

The graph of $y = \ln x$ has the same general shape as other logarithmic functions. It is increasing, has domain $(0, \infty)$, range $(-\infty, \infty)$, and a vertical asymptote at $x = 0$.

Example 11: Expand the logarithmic expression

$$\log(x^2(x-1)^3).$$

Solution:

$$\log x^2 + \log(x-1)^3 = 2\log x + 3\log(x-1).$$

Example 12: Given the logarithmic function

$$f(x) = \log_2(x-3) + 1,$$

(a) Find the domain.
(b) Identify the vertical asymptote.
(c) Find the x-intercept.
(d) Determine if the function is increasing or decreasing.
(e) Describe the transformations from $y = \log_2 x$.

Solution:

(a) The domain requires $x - 3 > 0$, so $x > 3$.
(b) Vertical asymptote: $x = 3$.
(c) To find the x-intercept:

$$\begin{aligned} 0 &= \log_2(x-3) + 1 \\ -1 &= \log_2(x-3) \\ x - 3 &= 2^{-1} = \frac{1}{2} \\ x &= \frac{7}{2}. \end{aligned}$$

(d) Since base $2 > 1$, the function is increasing.
(e) Transformations: shift right 3 units and up 1 unit.

For large values of x, the graph of $f(x) = \log_a x$ increases slowly and continues to grow without bound, but at a decreasing rate.

Group Exercises 16

Logarithmic Functions and Equations

(1) Rewrite each logarithmic equation in exponential form and solve.

(a) $\log_3 x = 4$

(b) $\log_5(x-2) = 2$

(c) $\ln x = 3$

(2) Expand each logarithmic expression.

(a) $\log(x^2y^3)$

(b) $\log\left(\frac{x^3}{y^2}\right)$

(c) $\ln(x^2(x-1)^4)$

(3) Condense each logarithmic expression into a single logarithm.

(a) $\log x + \log(x-2)$

(b) $2\log x - \log y$

(c) $3\ln x + 2\ln(x+1)$

(4) For each logarithmic function:

(i) Find the domain.
(ii) Identify the vertical asymptote.
(iii) Determine if it is increasing or decreasing.
(iv) Find the x-intercept.

(a) $f(x) = \log_2 x$

(b) $g(x) = \log_3(x-1)$

(c) $h(x) = \ln(x+4)$

(5) Sketch each function on the grids below. Label asymptotes and intercepts.

(a) $f(x) = \log_2(x - 2) + 1$

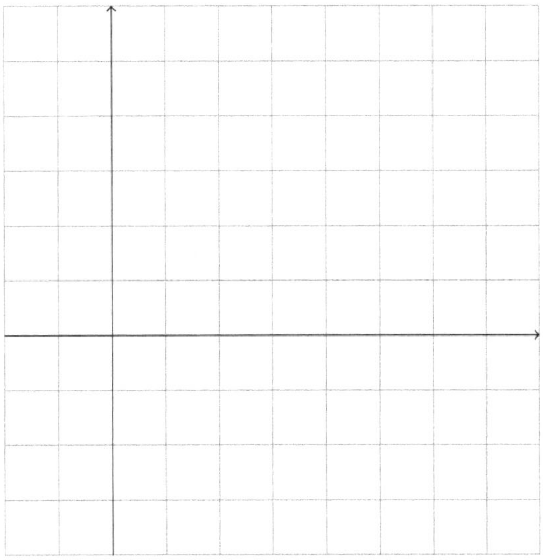

(b) $g(x) = \log_3(x + 1) - 2$

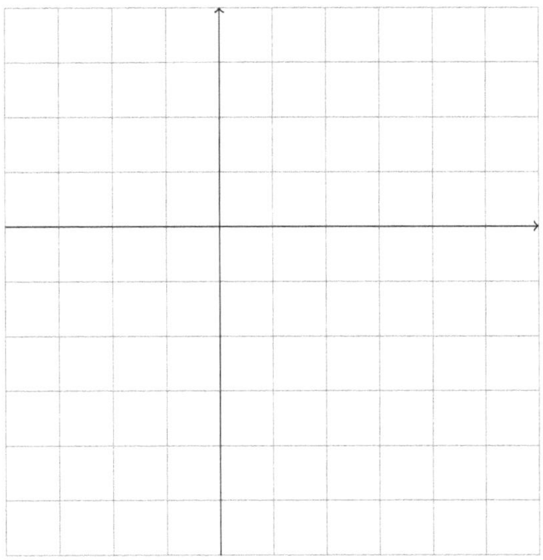

(c) $h(x) = -\ln(x-3)$

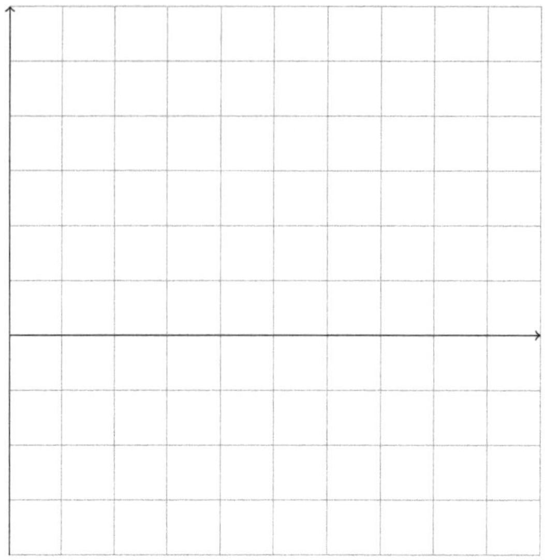

(6) Solve each equation:

(a) $\log_3 x = 2$

(b) $2^x = 10$

(c) $2^{x+1} = 5^{1-2x}$

(d) $\ln e^{-2x} = 8$

(e) $\log_2(x+7) + \log_2(x+8) = 1$

(f) $2\log_3(x+4) - \log_3 9 = 2$

Practice Exercises 16

Logarithmic Functions and Equations

(1) Rewrite each logarithmic equation in exponential form and solve.

(a) $\log_2 x = 5$

(b) $\log_4(x+1) = 3$

(c) $\ln x = 2$

(2) Expand each logarithmic expression.

(a) $\log(x^3y^2)$

(b) $\log\left(\frac{x^4}{y^3}\right)$

(c) $\ln(x^3(x+2)^2)$

(3) Condense each logarithmic expression into a single logarithm.

(a) $\log x + \log(x+3)$

(b) $3\log x - \log y$

(c) $2\ln x + \ln(x-1)$

(4) For each logarithmic function:
(i) Find the domain.
(ii) Identify the vertical asymptote.
(iii) Determine if it is increasing or decreasing.
(iv) Find the x-intercept.

(a) $f(x) = \log_5 x$

(b) $g(x) = \log_2(x+2)$

(c) $h(x) = \ln(x-3)$

(5) Sketch each function on the grids below. Label asymptotes and intercepts.

(a) $f(x) = \log_2(x+1) - 1$

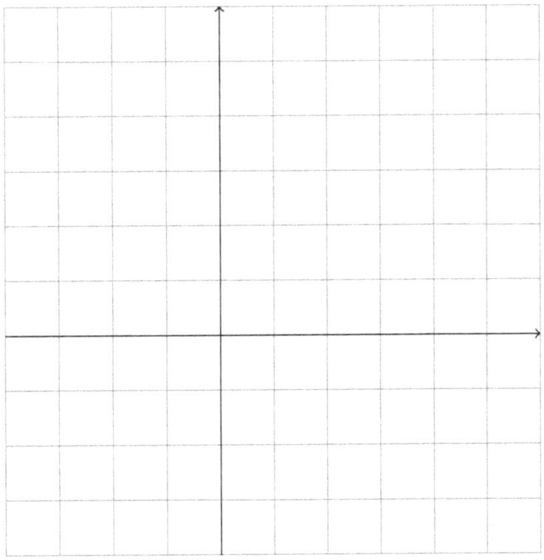

(b) $g(x) = \log_4(x-3) + 2$

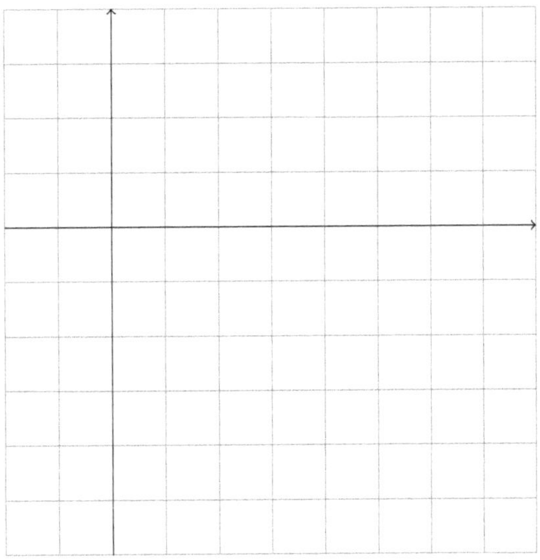

(c) $h(x) = -\ln(x+2)$

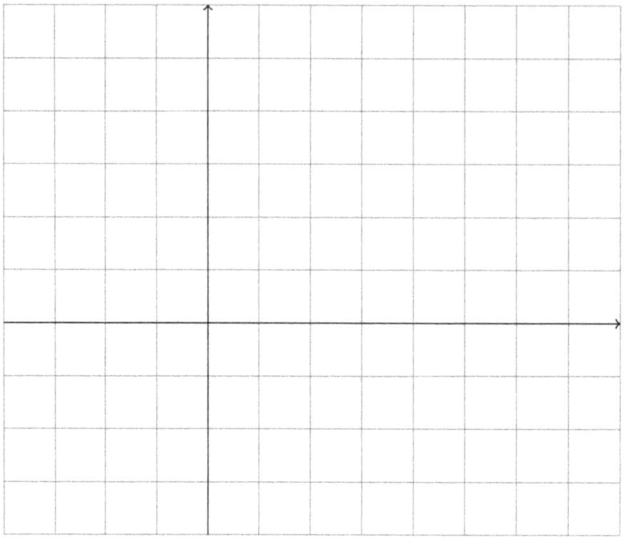

(6) Solve each equation:

(a) $\log_4 x = 3$

(b) $3^x = 7$

(c) $2^{x-1} = 4^{2x}$

(d) $\ln e^{3x} = -6$

(e) $\log_3(x+2) + \log_3(x+5) = 2$

(f) $2\log_2(x+1) - \log_2 8 = 1$

Chapter 17

Systems of Linear Equations

A system of linear equations is a set of two or more linear equations involving the same variables. A solution is an ordered pair that satisfies all equations simultaneously.

Types of Systems:

Type	Description
Independent	One solution (lines intersect at one point)
Inconsistent	No solution (parallel lines)
Dependent	Infinitely many solutions (same line)

Methods for Solving Systems:

- Substitution
- Elimination (Addition Method)
- Graphing (Visualization)

Example 1: (Substitution – One Solution). Solve : $\begin{cases} y = x + 2 \\ 2x + y = 5 \end{cases}$

Solution: Substitute $y = x + 2$ in the second equation and simplify:

$$2x + (x + 2) = 5 \Rightarrow 3x = 3 \Rightarrow x = 1$$

Now, substitute the value of x in the first equation

$$y = 1 + 2 = 3.$$

So the solution is the order pair: $\boxed{(1, 3)}$

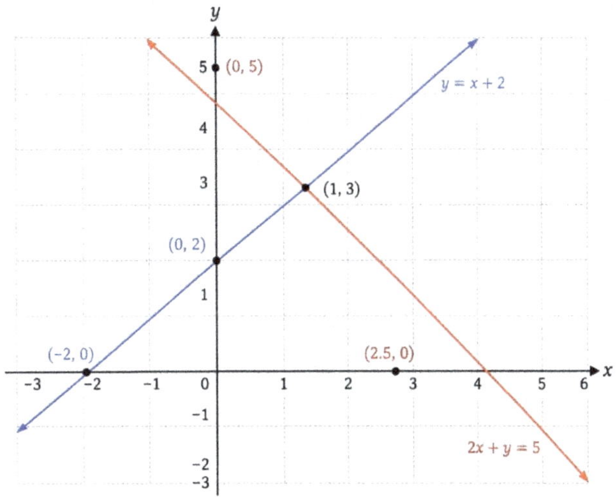

Example 2 (Elimination – One Solution). Solve : $\begin{cases} x + y = 4 \\ x - y = 2 \end{cases}$

Solution: Add both equations:

$$2x = 6 \Rightarrow x = 3$$

Now, substitute the value of x in the first equation, and solve for y:

$$y = 4 - 3 = 1.$$

So, the solution is the order pair: $\boxed{(3, 1)}$

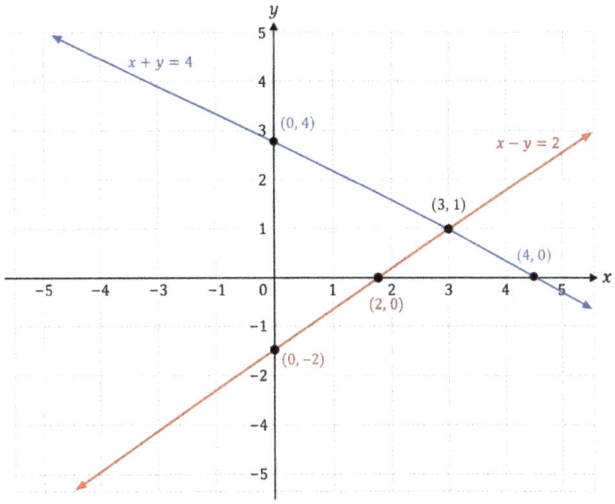

Example 3: (No Solution – Inconsistent System). Solve : $\begin{cases} y = 2x + 1 \\ y = 2x - 3 \end{cases}$

Solution: Same slope, different intercepts $\Rightarrow$ parallel lines.

$$\boxed{\text{No solution}}$$

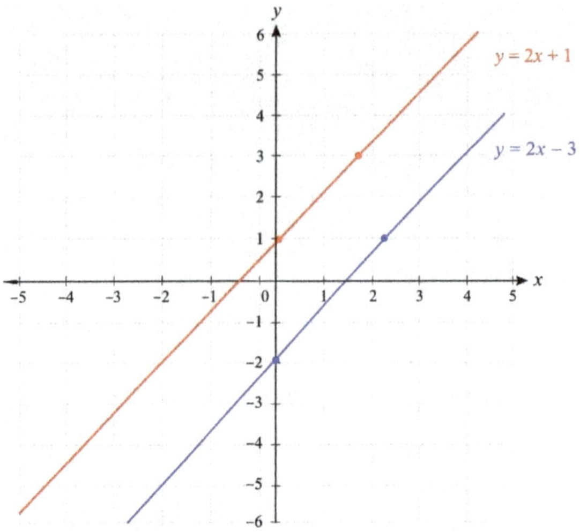

Example 4: (Infinitely Many Solutions — Dependent System). Solve : $\begin{cases} 2x + 4y = 8 \\ x + 2y = 4 \end{cases}$

Solution: Divide first equation by 2: $x + 2y = 4$. Now the first and second equation are identical, this implies that there are

$$\boxed{\text{Infinitely many solutions}}$$

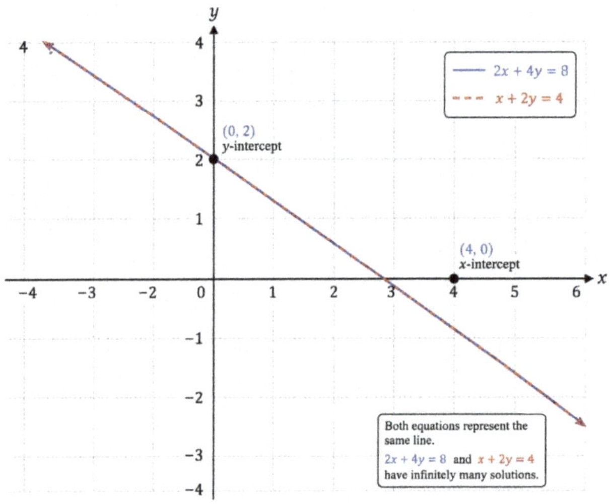

Problem I: (Word Problem — Fencing a Rectangular Field). A farmer wants to enclose a rectangular field using 100 meters of fencing. The length is 10 meters longer than the width. Find the dimensions of the field.

Solution:
Let:

$$w = \text{width}, \quad l = \text{length}$$

The perimeter of the rectangle formula:

$$2l + 2w = 100$$

Since the length is 10 meters longer than the width:

$$l = w + 10.$$

Write the system using the two equations:

$$\begin{cases} 2l + 2w = 100 \\ l = w + 10 \end{cases}$$

Substitute the second equation in the first equation and solve for w:

$$2(w + 10) + 2w = 100$$

$$2w + 20 + 2w = 100 \Rightarrow 4w = 80 \Rightarrow w = 20.$$

Now find l using the equation $l = w + 10$:

$$l = w + 10 = 20 + 10 = 30$$

$$\boxed{\text{Width } = 20 \text{ m, Length } = 30 \text{ m}}$$

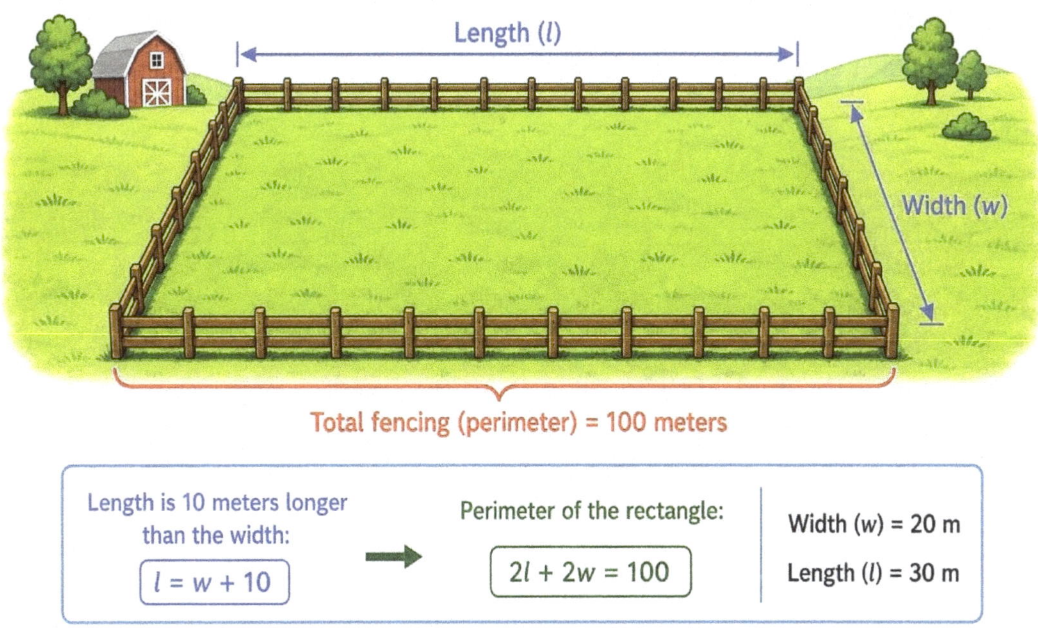

Summary:

- Systems of linear equations can have one, none, or infinitely many solutions.
- Substitution is useful when a variable is isolated.
- Elimination is efficient for combining equations.
- Graphs visually represent the number of solutions.
- Systems are widely used in real-world applications such as geometry and finance.

Problem II: A theater sold tickets for a school play. Adult tickets cost \$12 each and student tickets cost \$7 each. A total of 180 tickets were sold, and the total revenue was \$1590.

(1) Let a represent the number of adult tickets sold and s represent the number of student tickets sold. Write a system of linear equations.

(2) Solve the system.

(3) How many adult tickets and student tickets were sold?

Solution:

(1) The system is

$$\begin{cases} a + s = 180 \\ 12a + 7s = 1590 \end{cases}$$

(2) Solve using substitution.

From the first equation:

$$s = 180 - a$$

Substitute into the second equation:

$$12a + 7(180 - a) = 1590$$

Distribute:

$$12a + 1260 - 7a = 1590$$

Combine like terms:

$$5a + 1260 = 1590$$

Subtract 1260:

$$5a = 330$$

Divide by 5:

$$a = 66$$

Now find s:

$$s = 180 - a = 180 - 66 = 114.$$

(3) Therefore, the theater sold

$$\boxed{\text{66 adult tickets}}$$

and

$$\boxed{\text{114 student tickets}}.$$

Group Exercises 17

Systems of Linear Equations

Part A: Substitution (One Solution)
Solve each system using substitution.

(1) $\begin{cases} y = x + 3 \\ 2x + y = 7 \end{cases}$

(2) $\begin{cases} y = 2x - 1 \\ x + y = 5 \end{cases}$

(3) $\begin{cases} y = -x + 4 \\ 3x + y = 10 \end{cases}$

Part B: Elimination (One Solution)
Solve each system using elimination.

(4) $\begin{cases} x + y = 6 \\ x - y = 2 \end{cases}$

(5) $\begin{cases} 2x + y = 9 \\ 2x - y = 1 \end{cases}$

(6) $\begin{cases} 3x + 2y = 12 \\ 3x - 2y = 4 \end{cases}$

Part C: Identify the Type of System

State whether each system is: Independent, Inconsistent, or Dependent

(7) $\begin{cases} y = 3x + 1 \\ y = 3x - 4 \end{cases}$

(8) $\begin{cases} 2x + 4y = 8 \\ x + 2y = 4 \end{cases}$

(9) $\begin{cases} y = -x + 2 \\ y = x - 2 \end{cases}$

Part D: Graphing (Sketch the Systems)

Graph each system and state the number of solutions.

(10) $\begin{cases} y = x + 1 \\ y = -x + 5 \end{cases}$

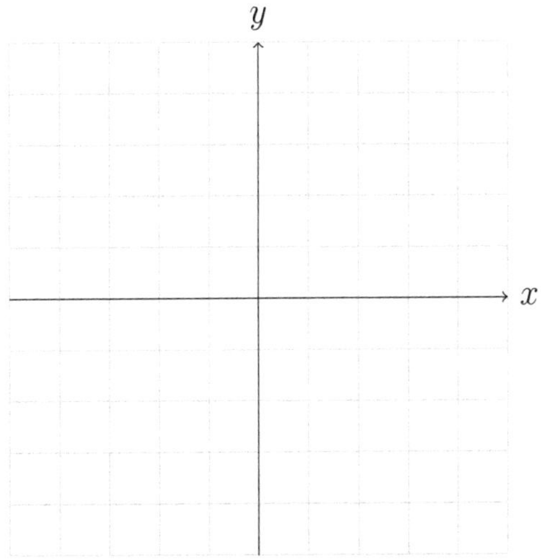

(11) $\begin{cases} y = 2x + 3 \\ y = 2x - 2 \end{cases}$

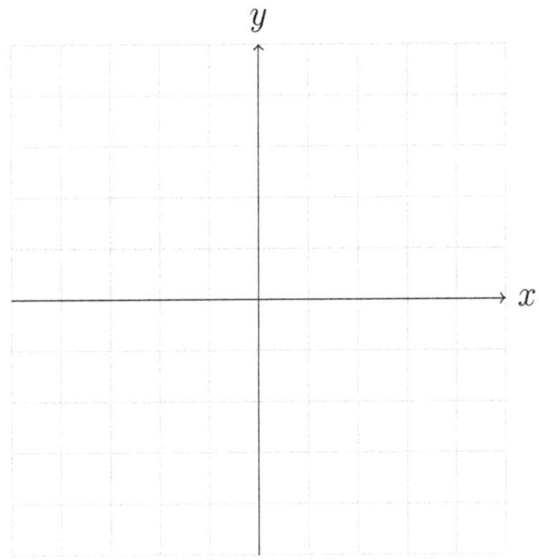

Part E: Word Problems (Group Discussion)

(12) A rectangle has a perimeter of 80 meters. The length is 8 meters longer than the width. Find the dimensions.

(13) The sum of two numbers is 20. One number is 4 more than the other. Find the two numbers.

(14) A school is buying pencils and notebooks. A pencil costs \$1 and a notebook costs \$3. If a student buys 10 items and spends \$22, how many of each did they buy?

Practice Exercises 17

Systems of Linear Equations

Part A: Substitution (One Solution)
Solve each system using substitution.

(1) $\begin{cases} y = x + 4 \\ x + y = 10 \end{cases}$

(2) $\begin{cases} y = 3x - 2 \\ 2x + y = 8 \end{cases}$

(3) $\begin{cases} y = -2x + 5 \\ x + y = 2 \end{cases}$

Part B: Elimination (One Solution)
Solve each system using elimination.

(4) $\begin{cases} x + y = 7 \\ x - y = 1 \end{cases}$

(5) $\begin{cases} 2x + y = 11 \\ 2x - y = 3 \end{cases}$

(6) $\begin{cases} 4x + y = 9 \\ 4x - y = -3 \end{cases}$

Part C: Identify the Type of System

State whether each system is: Independent, Inconsistent, or Dependent

(7) $\begin{cases} y = 2x + 3 \\ y = 2x - 5 \end{cases}$

(8) $\begin{cases} 3x + 6y = 12 \\ x + 2y = 4 \end{cases}$

(9) $\begin{cases} y = x + 1 \\ y = -x + 5 \end{cases}$

Part D: Graphing (Sketch the Systems)

Graph each system and state the number of solutions.

(10) $\begin{cases} y = x + 2 \\ y = -x + 4 \end{cases}$

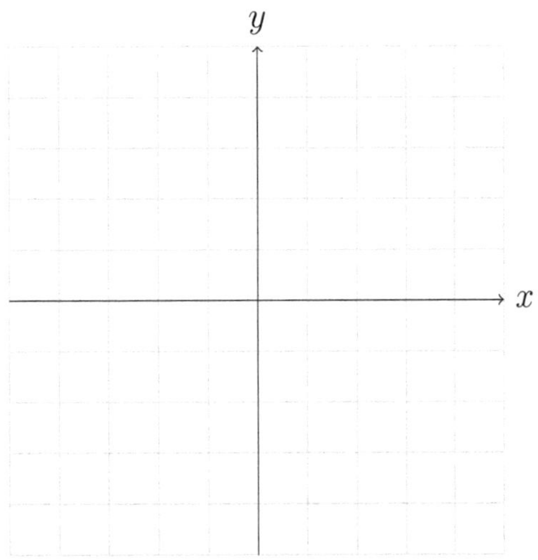

(11) $\begin{cases} y = 3x + 1 \\ y = 3x - 3 \end{cases}$

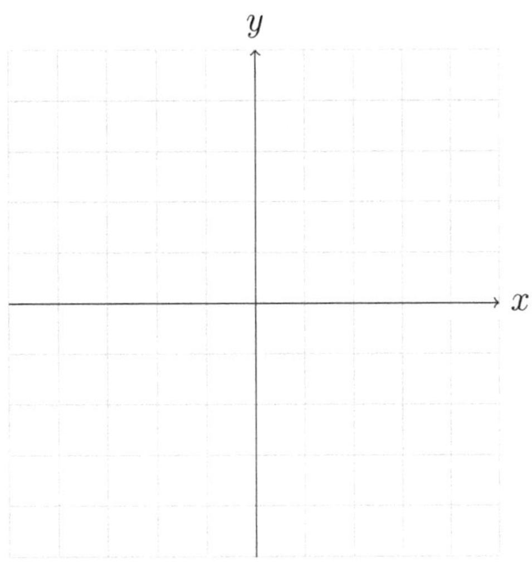

Part E: Word Problems (Group Discussion)

(12) A rectangle has a perimeter of 60 meters. The length is 6 meters longer than the width. Find the dimensions.

(13) The sum of two numbers is 26. One number is 8 more than the other. Find the two numbers.

(14) A school store sells pens for \$2 each and folders for \$4 each. If a student buys 9 items and spends \$24, how many of each did they buy?

(15) A concession stand sold hamburgers and hot dogs during a football game. Hamburgers cost \$5 each and hot dogs cost \$3 each. A total of 250 items were sold, and the total revenue was \$970.

(a) Let h represent the number of hamburgers sold and d represent the number of hot dogs sold. Write a system of linear equations.

(b) Solve the system.

(c) How many hamburgers and hot dogs were sold?

Appendix A: Instructions for Group Projects

A group project consists of a small group of three to five students. The project should be submitted as a written report by the due date and should consist of the following:

(1) Cover page (creative design by students)

(2) Title page (project title, date, students' names)

(3) Table of contents

(4) Summary (a one-page abstract of the problem, the approach used, and the results obtained)

(5) Supporting data (computations, labeled drawings, labeled computer plots, and/or printouts)

(6) Group log (time, date, location, and a brief description of each meeting)

(7) Evaluation summary of the group's learning experience while working on the project

(8) References (a list of references consulted)

All group members should be involved in answering each of the questions. In addition, each member of the group should be assigned a particular responsibility related to the project, such as Leader, Recorder, Checker, Typist, or Reader.

The project may require students to conduct surveys, research a topic, interview officials, compare prices, and compile their work into a formal report. The instructor may conduct individual interviews with group members and ask questions to ensure that all students are involved in answering every part of the project.

Project 1

"Line of Sight in the Rocky Mountains"

Introduction: Students will model a mountain skyline given by

$$y = -x^4 - 2x^3 + 12x^2 + x - 10, \quad -4 < x < 3,$$

and determine a line of sight that just grazes the two peaks. This represents a path that touches the highest points without cutting through the mountain.

Project Goals:

- Graph polynomial functions
- Identify local maxima
- Estimate and refine linear equations
- Interpret results in context

Problem: Find the equation of a line that just grazes the two peaks of the given skyline.

Tasks:

(1) Assign a Plotter to graph the skyline.
(2) Estimate the coordinates of the two peaks.
(3) Take turns guessing a line $y = mx + b$.
(4) Graph each guess and evaluate its accuracy.
(5) Describe errors (too steep, too low, intersects curve, etc.).
(6) Refine guesses until the line just touches both peaks.
(7) Agree on criteria for a "suitable" line.
(8) State the final equation.
(9) Briefly explain how your group improved its guesses.
(10) Interpret the result in terms of real-world applications.

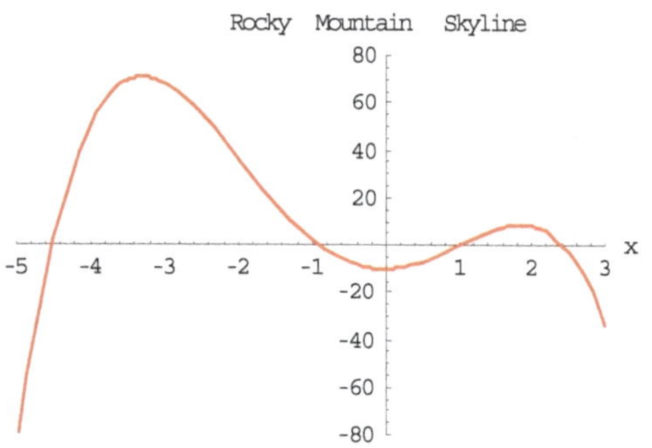

Project 2

"Comparing Transportation Costs"

Introduction: A group of students is deciding how to commute to campus each day. Several transportation options are available, each with different pricing structures. Understanding these options requires modeling costs using linear functions and analyzing graphs and systems of equations. The goal is to determine which option is most economical under different circumstances.

The three transportation options are:

a) Ride Service: A base fee of \$3 per trip plus \$0.50 per mile
b) Car Rental: A flat daily fee of \$25 plus \$0.20 per mile
c) Public Transit: A flat fee of \$10 per day for unlimited travel

Your group's task is to compare the three options for daily travel distances of 0, 5, 10, 20, 30, and 50 miles by doing the following:

(1) Construct a table of values showing the total daily cost for each transportation option at the given distances.

(2) Draw (by hand) the graph of cost versus distance for each option. Plot all three graphs on the same coordinate system. Let distance be measured on the horizontal axis and cost on the vertical axis.

(3) Use your graph to estimate the distance(s) at which the cost of two options is approximately the same.

(4) Write a cost function for each option in terms of distance. Let the independent variable represent distance traveled and the dependent variable represent cost.

(5) Use algebra to find the exact point(s) of intersection of the cost functions. Compare your answers with your estimates from the graph.

(6) Suppose a student travels an average of 15 miles per day. Which transportation option is the most economical? What if the student travels 40 miles per day? Explain your reasoning.

(7) Discuss the advantages and disadvantages of each transportation option. Consider not only cost, but also convenience, flexibility, and predictability.

Project 3

"Cell Phone Plans"

Introduction: A group of students is trying to decide which cell phone plan gives the best value. Several companies offer different pricing methods, and consumers must compare fixed monthly fees, usage charges, and data limits in order to choose the most economical plan. This project uses algebraic models, tables, graphs, and systems of equations to analyze and compare several cell phone plans.

Project Goals: This project helps students practice:

- Writing and interpreting linear functions
- Constructing tables of values
- Graphing multiple linear models on the same axes
- Solving systems of linear equations by finding intersection points
- Interpreting algebraic results in a real-world setting
- Comparing mathematical models and making decisions based on data

Suppose three companies offer the following monthly plans:

a) Plan A: A monthly fee of \$25 plus \$0.10 per text message
b) Plan B: A monthly fee of \$40 plus \$0.05 per text message
c) Plan C: A flat monthly fee of \$55 for unlimited text messages

Your group's task is to compare the three plans and determine which plan is the best choice for different types of customers.

(1) Construct a table of values showing the monthly cost for each plan if a customer sends 0, 100, 200, 300, 400, and 500 text messages.

(2) On the same coordinate axes, graph the cost of each plan as a function of the number of text messages sent per month. Let the horizontal axis represent the number of text messages and the vertical axis represent the monthly cost in dollars.

(3) Write a cost function for each plan. Let the independent variable represent the number of text messages sent in a month, and let the dependent variable represent the monthly cost.

(4) Use algebra to determine the point(s) of intersection of the graphs. Interpret each intersection point in practical terms.

(5) Based on your graph and equations, determine for what range of text-message usage each plan is the least expensive option.

(6) Suppose a customer usually sends about 250 text messages per month. Which plan should the customer choose? What if the customer sends about 450 text messages per month? Explain your reasoning.

(7) Many people prefer a plan with a predictable monthly cost, even if it is not always the cheapest. Discuss the advantages and disadvantages of each plan from the perspective of a consumer.

(8) Create a new Plan D of your own. Your group should design a pricing plan that you believe would be attractive to customers but still profitable for the company. Write its cost function, graph it with the other three plans, and explain which types of customers would benefit most from your new plan.

Group Exercises Answer Key

Group Exercises 1: Linear Equations and Applications

(1) (a) $x = 1$ (b) $a = -12$ (c) $x = -1$ (d) $p = 2$ (e) $y = -\frac{1}{2}$ (f) no solution (g) $x = 34$.

(2) (a) $x = \frac{b+c}{a}$, $a \neq 0$. (b) $w = \frac{l}{t+1}$, $t \neq -1$. (c) $x = x_1 + \frac{y-y_1}{m}$, $m \neq 0$.

(3) Yes; they intersect at $(-3, 5)$.

(4) $\frac{50}{65} \cdot 60 = \frac{600}{13} \approx 46.15$ minutes.

(5) (a) $A(x) = 0.15x + 1$ (b) $B(x) = 0.25x$ (c) $A(45) = \$7.75$, $B(45) = \$11.25$ (d) Company A: $26\frac{2}{3}$ min; Company B: 20 min (e) $x = 10$ min.

Group Exercises 2: Quadratic Equations and Applications

Factoring: (1) $x = \pm 3$ (2) $x = 0, \frac{1}{2}$ (3) $x = -3, 2$ (4) $x = 3, -\frac{1}{2}$ (5) $x = \frac{3}{2}$.
Completing square: (1) $x = -7, 3$ (2) $x = 3 \pm \sqrt{22}$ (3) $x = 1, -4$.
Quadratic formula: (1) $x = 2 \pm \sqrt{2}$ (2) $x = 1, \frac{3}{2}$ (3) $x = -3 \pm 2\sqrt{2}$ (4) no real solutions.
Nature of roots: (1) two complex conjugate roots (2) two distinct real roots (3) one repeated real root.
Problem I: width 17 cm, length 18 cm.
Problem II: (1) $t = 6$ s (2) $t = 5$ s.

Group Exercises 3: Radical Equations and Factoring Methods

(1) $t = 1$.

(2) No real solution.

(3) $x = 3$.

(4) $x = -13$.

(5) $x = 0, 1$.

(6) $x = 1, 5$.

Substitution equations: (1) let $u = x^2$; $x = \pm\sqrt{5}$. (2) let $u = 1 - y$; $y = \frac{4}{3}, 2$.
Factoring: (1) $x = 0, \frac{3}{4}$ (2) $x = 0$ only as a real solution (3) $x = -1, 1, 3$.

Group Exercises 4: Inequalities and Interval Notation

(1) (a) $[0, 4]$ (b) $(-4, 1]$ (c) $[-5, 1)$ (d) $(3, 7)$ (e) $(-\infty, 13)$ (f) $(1, \infty)$ (g) $(-\infty, -2]$ (h) $[0, \infty)$.

(2) (a) $2 \leq x \leq 5$ (b) $-2 < x < -1$ (c) $0 < x \leq 5$ (d) $-3 \leq x < 0$ (e) $x \geq -7$ (f) $x > 2$ (g) $x \leq 1$ (h) $x < -3$.

(3) (a) $x \geq -1$ (b) $x \leq \frac{2}{3}$ (c) $x \geq 5$ (d) $x < -20$ (e) $-\frac{11}{2} < x < \frac{1}{2}$ (f) $\frac{1}{2} \leq x < \frac{5}{4}$ (g) $x < -\frac{1}{2}$.

Problem I: if s is the last test score, $74 \leq s < 124$; with a usual 0–100 scale, $74 \leq s \leq 100$.
Problem II: usage ranges from $\frac{63.47-7.58}{0.08275} \approx 675.41$ kWh to $\frac{214.53-7.58}{0.08275} \approx 2500.91$ kWh.

Group Exercises 5: Solving Absolute Value Equations and Inequalities

(1) (a) $x = -\frac{1}{2}, \frac{3}{2}$ (b) $t = -1, \frac{3}{2}$ (c) no solution (d) $x = \pm 4$ (e) $x = -\frac{4}{3}, \frac{8}{3}$.

(2) (a) $[-1, 3]$ (b) $(-6, -2)$ (c) $(-1, 2)$ (d) $(-\infty, -6] \cup [-2, \infty)$ (e) $(-\infty, \frac{1}{3}) \cup (1, \infty)$ (f) $[-1, 2]$.

Problem I: $|x - 98.6| \geq 1.5$, so $x \leq 97.1$ or $x \geq 100.1$.

Group Exercises 6: Lines and Their Graphs

(1) Plot: $A(-3,2), B(0,3), C(5,0), D(-3,-2), E(-1,0)$.
(2) Distances: (a) $2\sqrt{17}$ (b) $2\sqrt{2}$ (c) $\sqrt{17}$ (d) 5.
(3) (a) Right triangle; area $\frac{13}{2}$. (b) Right triangle; area 10.
(4) Midpoints: (a) $(4,-1)$ (b) $(5,-1)$ (c) $(-\frac{1}{2},-2)$ (d) $(\frac{7}{2},\frac{5}{2})$.
(5) Slopes: (a) $-\frac{3}{2}$ (b) undefined (c) -1.
(6) (a) line through $(1,2)$ with slope 3: e.g. points $(1,2),(2,5),(0,-1)$. (b) vertical line $x=0$.
(7) (a) Additional points may be $(-1,5),(0,7)$. (b) Additional points may be $(-1,-5),(0,-7)$.

Group Exercises 7: Equations of Lines and Applications

(1) (a) $y=\frac{1}{3}x+\frac{4}{3}$ (b) $y=-\frac{1}{2}x$.
(2) (a) $2x-y+5=0$ (b) $x-1=0$ (c) $3x+y=0$ (d) $x-y-4=0$ (e) $y-2=0$.
(3) (a) $m=3$, $b=-1$ (b) $m=\frac{2}{5}$, $b=\frac{3}{5}$ (c) $m=-\frac{3}{4}$, $b=-\frac{1}{2}$ (d) $m=0$, $b=4$ (e) slope undefined; no y-intercept.
(4) (a) parallel 0, perpendicular undefined (b) parallel undefined, perpendicular 0 (c) parallel $-\frac{3}{5}$, perpendicular $\frac{5}{3}$ (d) parallel -3, perpendicular $\frac{1}{3}$.
(5) $y=2x$.

Problem I: $F=\frac{9}{5}C+32$; when $F=70$, $C=\frac{190}{9}\approx 21.11°\text{C}$.
Problem II: $S=C+5$; a book that sells for 52 costs 47.

Group Exercises 8: Functions: Notation, Graphs, Domain, and Range

(1) (a) Not a function (b) function.
(2) (a) -4 (b) -3 (c) 1 (d) $3x^2-2x-4$ (e) $3x^2+6xh+3h^2+2x+2h-4$.
(3) (a) function (b) function (c) function.
(4) (a) yes (b) no (c) no (d) yes.
(5) (a) $(-\infty,\frac{3}{2}]$ (b) $\mathbb{R}\setminus\{-4,4\}$ (c) $(-5,\infty)$ (d) $\mathbb{R}\setminus\{0,4\}$.
(6) (a) domain $\mathbb{R}$, range $\mathbb{R}$ (b) domain $\mathbb{R}$, range $[0,\infty)$ (c) domain $[1,\infty)$, range $[0,\infty)$.
(7) (a) $(f+g)(x)=5x+1$ (b) $(f-g)(x)=x+7$ (c) $(fg)(x)=6x^2-x-12$ (d) $\left(\frac{f}{g}\right)(x)=\frac{3x+4}{2x-3}$, $x\neq\frac{3}{2}$.

Group Exercises 9: Graphs of Functions

(1) (a) No (b) $f(-2)=0$ (c) $x=2$.
(2) (a) $f(0)=3$, $f(-6)=-3$, $f(6)=0$, $f(11)=1$ (b) $x=-3,6,10$ (c) $(-3,0),(6,0),(10,0)$ (d) $(0,3)$ (e) negative (f) positive.
(3) (a) odd (b) odd (c) neither (d) neither.
(4) For $f(x)=-2x+4$: (a) -2 (b) -2 (c) -2.

Group Exercises 10: Library of Functions

(1) (a) $D=\mathbb{R}$, $R=\mathbb{R}$. (b) $D=\mathbb{R}$, $R=(-\infty,4]$. (c) $D=\mathbb{R}$, $R=[0,\infty)$. (d) $D=\mathbb{R}$, $R=(-\infty,2]$.
(2) (a) 4 (b) 2 (c) 5 (d) $D=\mathbb{R}$ (e) $R=(0,\infty)$.

Problem I: (a) \$39.99 (b) \$43.74 (c) \$40.24.

Group Exercises 11: Graphing Techniques and Transformations

(1) Key transformations: (a) x^2 shifted down 2; vertex $(0,-2)$. (b) x^3 shifted down 1; inflection $(0,-1)$. (c) $\sqrt{x}$ shifted left 2; starts at $(-2,0)$. (d) x^2 shifted right 1, stretched by 3, up 1; vertex $(1,1)$. (e) $|x|$ shifted left 1, reflected over x-axis, up 3; vertex $(-1,3)$. (f) $1/x$ reflected over the y-axis and shifted up 2; asymptotes $x=0$, $y=2$. (g) x^2 vertically compressed by $\frac{1}{2}$ and shifted down 2; vertex $(0,-2)$.

(2) (a) $y=\sqrt{x}+2$ (b) $y=-\sqrt{x}+2$ if applied after (a); otherwise $y=-\sqrt{x}$ (c) $y=\sqrt{-x}$.

(3) (a) $R(x)=x(-\frac{1}{6}x+100)=-\frac{1}{6}x^2+100x$ (b) $R(200)=\$13,333.33$ (c) downward-opening parabola (d) $x=300$ units; maximum revenue $\$15,000$.

Group Exercises 12: Quadratic Functions and Models

(1) (a) $(x-3)^2+1$ (b) $2(x+\frac{3}{2})^2+\frac{1}{2}$ (c) $-(x+\frac{1}{2})^2+\frac{1}{4}$.

(2) (a) vertex $(1,-2)$; intercepts $(0,0),(2,0)$ and $(0,0)$. (b) vertex $(\frac{1}{2},0)$; x-intercept $(\frac{1}{2},0)$; y-intercept $(0,-1)$. (c) vertex $(-3,-21)$; y-intercept $(0,-3)$; x-intercepts $\left(\frac{-6\pm\sqrt{42}}{2},0\right)$.

(3) From each graph, read the vertex and one additional point, then use $f(x)=a(x-h)^2+k$.

Problem I: price $p=500$; maximum revenue $\$1,000,000$.
Problem II: maximum at $t=\frac{5}{2}$ seconds; maximum height 100 ft.

Group Exercises 13: Polynomial Functions

(1) (a) not polynomial (variable in denominator) (b) polynomial, degree 1 (c) polynomial, degree 5 (d) not polynomial (variable in denominator).

(2) (a) $(x+1)(x-1)(x-3)$ (b) $(x+4)(x+1)(x-2)(x-3)$ (c) $(x+1)(x-3)^2$.

(3) (a) zeros 7 mult. 1 crosses, -3 mult. 2 touches; end behavior like $3x^3$. (b) zeros 5 mult. 3 crosses, -4 mult. 2 touches; like x^5. (c) zeros 0 mult. 4 touches, 2 mult. 1 crosses; like $-2x^5$.

(4) (a) x-intercepts 0 (touch), 3 (cross); y-intercept 0; like x^3; max turns 2; below $(-\infty,0)\cup(0,3)$, above $(3,\infty)$, and touches at $x=0$. (b) x-intercepts $-4,1,2$ all cross; y-intercept -8; like x^3; max turns 2; below $(-\infty,-4)\cup(1,2)$, above $(-4,1)\cup(2,\infty)$. (c) x-intercepts -2 cross, 0 touch, 2 cross; y-intercept 0; like x^4; max turns 3; above $(-\infty,-2)\cup(2,\infty)$, below $(-2,0)\cup(0,2)$.

Group Exercises 14: Rational Functions and Asymptotes

(1) (a) $x\neq 3$ (b) $x\neq 3,-\frac{1}{2}$ (c) domain $\mathbb{R}$.

(2) (a) VA $x=-4$, HA $y=3$, no oblique. (b) VA $x=\pm1$, HA $y=0$, no oblique. (c) VA $x=0$, no HA, oblique asymptote $y=3x$. (d) VA $x=\pm5$, no HA, oblique asymptote $y=x+4$.

(3) (a) VA $x=3$, HA $y=1$, range $(1,\infty)$. (b) VA $x=-1$, HA $y=2$.

Problem I: (a) HA $C=0$; concentration approaches 0. (b) graph is positive for $t>0$, rises then falls toward 0. (c) maximum at $t=5$ minutes, with $C(5)=5$.

Group Exercises 15: Exponential Functions and Equations

(1) (a) exponential, base 2 (b) not exponential; variable is the base, not exponent (c) exponential, base $\frac{1}{3}$ (d) not an exponential function over all real x because the base is negative (e) exponential, base e.

(2) (a) $2^7=128$ (b) $5^4=625$ (c) $3^6=729$ (d) e^5 (e) e^3.

(3) (a) $x=4$ (b) $x=3$ (c) $x=\frac{15}{2}$ (d) $x=1,5$ (e) $x=\frac{1}{2}$ (f) $x=\frac{8}{3}$.

(4) (a) $D=\mathbb{R}$, $R=(-2,\infty)$, HA $y=-2$. (b) $D=\mathbb{R}$, $R=(-\infty,2)$, HA $y=2$. (c) $D=\mathbb{R}$, $R=(-\infty,1)$, HA $y=1$. (d) $D=\mathbb{R}$, $R=(-\infty,5)$, HA $y=5$.

Group Exercises 16: Logarithmic Functions and Equations

(1) (a) $x = 81$ (b) $x = 27$ (c) $x = e^3$.
(2) (a) $2\log x + 3\log y$ (b) $3\log x - 2\log y$ (c) $2\ln x + 4\ln(x-1)$.
(3) (a) $\log(x(x-2))$ (b) $\log\left(\frac{x^2}{y}\right)$ (c) $\ln(x^3(x+1)^2)$.
(4) (a) domain $(0, \infty)$, VA $x = 0$, increasing, x-intercept $(1, 0)$. (b) domain $(1, \infty)$, VA $x = 1$, increasing, x-intercept $(2, 0)$. (c) domain $(-4, \infty)$, VA $x = -4$, increasing, x-intercept $(-3, 0)$.
(5) (a) VA $x = 2$, x-intercept $x = \frac{5}{2}$. (b) VA $x = -1$, x-intercept $x = 8$. (c) VA $x = 3$, x-intercept $x = 4$.
(6) (a) $x = 9$ (b) $x = \log_2 10$ (c) $x = \frac{\ln(5/2)}{\ln(50}$ (d) $x = -4$ (e) no real solution (f) $x = 5$.

Group Exercises 17: Systems of Linear Equations

(1) $(\frac{4}{3}, \frac{13}{3})$.
(2) $(2, 3)$.
(3) $(3, 1)$.
(4) $(4, 2)$.
(5) $(\frac{5}{2}, 4)$.
(6) $(\frac{8}{3}, 2)$.
(7) Inconsistent.
(8) Dependent.
(9) Independent; one solution $(2, 0)$.
(10) One solution: $(2, 3)$.
(11) No solution.
(12) Width 16 m, length 24 m.
(13) Numbers 8 and 12.
(14) 4 pencils and 6 notebooks.

Practice Exercises Answer Key

Practice Exercises 1: Linear Equations and Applications

(1) $(-\frac{39}{7}, \frac{4}{7})$.

(2) (a) No solution.
  (b) $x = \frac{7}{5}$.
  (c) $x = \frac{12}{13}$.
  (d) $x = -12$.
  (e) $x = -1$.

(3) (a) $y = \frac{2x-7}{5}$.
  (b) $x = \frac{8}{2k-3}$, $k \neq \frac{3}{2}$.
  (c) $b = \frac{2A}{h} - a$, $h \neq 0$.

(4) (a) 57, 790.38 pesos.
  (b) 407, 527.29 yen.
  (c) 1, 986.15 dollars.

(5) (a) Jim: $J(x) = 20x + 50$; Jill: $K(x) = 490$.
  (b) Break-even: $x = 22$. Use Jim if $x < 22$, either if $x = 22$, and Jill if $x > 22$.

Practice Exercises 2: Quadratic Equations and Applications

(1) Factoring: $x = \pm 5$.
(2) $v = \pm 5$.
(3) $x = -1, 8$.
(4) $t = \frac{1}{3}, 1$.
(5) $x = -\frac{1}{2}, \frac{2}{3}$.
(6) Completing square: $t = -3, 5$.
(7) $x = -4, -1$.
(8) $x = \frac{3 \pm \sqrt{17}}{4}$.
(9) Quadratic formula: no real solutions.
(10) $x = 2 \pm \sqrt{5}$.
(11) $x = -1 \pm \frac{\sqrt{10}}{2}$.
(12) $x = 1, -\frac{2}{3}$.
(13) Nature of roots: two complex conjugate roots.
(14) Two distinct real roots.
(15) One repeated real root.
(16) Problem I: cliff height 64 ft; maximum height 100 ft; time to maximum $\frac{3}{2}$ s; reaches 32 ft at $t = \frac{3+\sqrt{17}}{2}$ s; hits ground at $t = 4$ s.
(17) Problem II: original sheet metal is 4 ft $\times$ 4 ft.

Practice Exercises 3: Radical Equations and Factoring Methods

(1) $t = 0$.
(2) No real solution.
(3) $x = 3$.
(4) $x = 0$.

(5) $x = 4$.
(6) $x = -\frac{5}{4}$.
(7) With $u = x^2$: $x = \pm 1, \pm 2$.
(8) With $u = s + 1$: $s = -\frac{3}{2}, 2$.
(9) $x = -1, 0, 1$.
(10) $x = -1, 1$.
(11) $x = -5, 0, 4$.
(12) $x = 0, \frac{3}{4}$.

Practice Exercises 4: Inequalities and Interval Notation

(1) Interval notation: (a) $[-10, -2]$; (b) $(1, 7]$; (c) $[0, 3)$; (d) $(-2, 2)$; (e) $(-\infty, 5)$; (f) $(-1, \infty)$; (g) $(-\infty, 0]$; (h) $[-4, \infty)$.
(2) Inequalities: (a) $-2 \leq x \leq 0$; (b) $-2 < x < 3$; (c) $-4 < x \leq 4$; (d) $-1 \leq x < 0$; (e) $x \geq 7$; (f) $x > -2$; (g) $x \leq -1$; (h) $x < 3$.
(3)
 (a) $x < 2$.
 (b) $x < 5$.
 (c) $x \leq 0$.
 (d) $x \geq -1$.
 (e) $a < \frac{9}{2}$.
 (f) $x \leq -\frac{1}{5}$.
 (g) $-\frac{2}{3} < x < 2$.
 (h) $-\frac{1}{3} < x \leq \frac{1}{3}$.
 (i) $x > \frac{1}{2}$.
(4) Problem I: more than 4 appliances, so at least 5 appliances.
(5) Problem II: minimum score 98.
(6) Problem III: commission ranges from \$2,000 to \$22,000.

Practice Exercises 5: Solving Absolute Value Equations and Inequalities

(1) Equations: (a) $x = 1, 3$; (b) $t = -1$; (c) $x = -3, \frac{3}{2}$; (d) $x = 0, -1$; (e) $x = -3, -1$.
(2) Inequalities: (a) $(-\frac{36}{5}, \frac{24}{5})$; (b) $[-\frac{2}{3}, 2]$; (c) $(-\frac{7}{2}, \frac{9}{2})$; (d) $(-\infty, -1] \cup [7, \infty)$; (e) $(-\infty, 0) \cup (0, \infty)$; (f) $(-\infty, \infty)$.
(3) Problem I: $|x - 115| \leq 5$, so $110 \leq x \leq 120$ volts.

Practice Exercises 6: Lines and Their Graphs

(1) Plot the listed points: $(-1, 0), (0, -1), (-3, -3), (2, 2), (\frac{1}{2}, \frac{3}{2})$.
(2) Distances: (a) $\sqrt{53}$; (b) $\sqrt{61}$; (c) $\sqrt{10}$; (d) $\sqrt{41}$.
(3) (a) Right triangle; area $\frac{29}{2}$. (b) Right triangle; area 6.
(4) Midpoints: (a) $(3, -\frac{1}{2})$; (b) $(-1, -\frac{1}{2})$; (c) $(\frac{3}{2}, 1)$; (d) $(\frac{a}{2}, \frac{b}{2})$.
(5) Slopes: (a) $-\frac{1}{2}$; (b) $-\frac{3}{2}$; (c) 0.
(6) Graphs: (a) line through $(2, 4)$ with slope $-\frac{3}{4}$; (b) horizontal line $y = -4$.
(7) Possible additional points: (a) $(3, 2), (2, 3)$; (b) $(0, 6), (3, 10)$.

Practice Exercises 7: Equations of Lines and Applications

(1) (a) $y = -\frac{1}{3}x + \frac{4}{3}$. (b) $y = \frac{1}{2}x$.

(2) (a) $x + 2y + 2 = 0$. (b) $x - 5y + 23 = 0$. (c) $2x + y = 0$. (d) $x - y + 4 = 0$. (e) $y + 21 = 0$.

(3) (a) $m = -3, b = 4$. (b) $m = -\frac{1}{2}, b = 2$. (c) $m = -\frac{3}{2}, b = 3$. (d) slope undefined, no y-intercept. (e) $m = 0, b = -3$.

(4) (a) parallel slope 0, perpendicular slope undefined. (b) parallel undefined, perpendicular 0. (c) parallel $\frac{1}{2}$, perpendicular -2. (d) parallel 6, perpendicular $-\frac{1}{6}$.

(5) $y = \frac{1}{2}x + \frac{3}{2}$.

(6) Problem I: $C = 29 + 0.07x$; $C(110) = \$36.70$.

(7) Problem II: $d = 18000 - 160p$; $d(50) = 10000$; decreasing price from 50 to 46 increases demand by 640; increasing price from 37.50 to 38 decreases demand by 80; no demand at $p = 112.50$.

Practice Exercises 8: Functions: Notation, Graphs, Domain, and Range

(1) (a) Not a function. (b) Function.

(2) $g(0) = -3$; $g(-3) = -6$; $g(3) = 0$; $g(-x) = -|-x-3| = -|x+3|$; $g(x+h) = -|x+h-3|$.

(3) From the graphs: use the vertical-line test; any graph with a vertical line intersecting more than once is not a function.

(4) (a) Function. (b) Not a function. (c) Not a function. (d) Function.

(5) (a) $(-\infty, -2]$. (b) $(9, \infty)$. (c) $(-\infty, -2) \cup (0, 2) \cup (2, \infty)$. (d) $(-\infty, \infty)$.

(6) Domain/range must be read from the three graphs.

(7) If $f(x) = \sqrt{x}$ and $g(x) = 3x - 5$: (a) $(f+g)(x) = \sqrt{x} + 3x - 5$; (b) $(f-g)(x) = \sqrt{x} - 3x + 5$; (c) $(fg)(x) = \sqrt{x}(3x-5)$; (d) $\left(\frac{f}{g}\right)(x) = \frac{\sqrt{x}}{3x-5}$, $x \geq 0$, $x \neq \frac{5}{3}$.

Practice Exercises 9: Graphs of a Functions

(1) For $g(x) = \frac{x^2+2}{x+4}$: (a) No, since $g(-1) = 1$. (b) $g(0) = \frac{1}{2}$. (c) $x = 0$; point $(0, \frac{1}{2})$.

(2) From the graph: (a) $f(0) = 0$, $f(-2) = 1$, $f(6) = 0$, $f(5) = 3$. (b) $x = 0, 4, 6$. (c) $(0,0), (4,0), (6,0)$. (d) $(0,0)$. (e) positive. (f) negative.

(3) (a) Neither. (b) Odd. (c) Even. (d) Odd.

(4) Average rates of change for $f(x) = -x^3 + 1$: (a) -4; (b) -13; (c) -1.

Practice Exercises 10: Library of Functions

(1) Domains/ranges: (a) $f(x) = 3x - 2$: domain $\mathbb{R}$, range $\mathbb{R}$. (b) $x^2 + 4$: domain $\mathbb{R}$, range $[4, \infty)$. (c) $|2x - 1|$: domain $\mathbb{R}$, range $[0, \infty)$. (d) piecewise: domain $[-4, \infty)$, range $[-1, \infty)$.

(2) For $h(x) = x^3$ if $x < 0$, and $h(x) = 3x + 2$ if $x \geq 0$: (a) -1; (b) 2; (c) 5; domain $\mathbb{R}$; range $(-\infty, 0) \cup [2, \infty)$.

(3) Problem I: (a) \$1.29; (b) \$0.39; (c) \$3.13.

Practice Exercises 11: Graphing Techniques and Transformations

(1) Graph keys: (a) x^2 shifted up 2, vertex $(0, 2)$. (b) x^3 shifted up 1, inflection $(0, 1)$. (c) $\sqrt{x}$ shifted right 1, start $(1, 0)$. (d) x^2 shifted left 2, vertically stretched by 2, down 1, vertex $(-2, -1)$. (e) $|x|$ shifted right 1, down 1, vertex $(1, -1)$. (f) $-1/x+2$, vertical asymptote $x = 0$, horizontal asymptote $y = 2$. (g) $-\frac{1}{2}x^2 + 1$, vertex $(0, 1)$.

(2) Check graph answers by listing the parent function, transformations, and key points/intercepts.

Practice Exercises 12: Quadratic Functions and Models

(1) Vertex form: (a) $-(x+1)^2-2$. (b) $3(x+2)^2+2$. (c) $-4\left(x-\frac{1}{2}\right)^2+1$.

(2) (a) Vertex $(-1,4)$; x-intercepts $(-3,0),(1,0)$; y-intercept $(0,3)$. (b) Vertex $(\frac{1}{2},\frac{3}{4})$; no real x-intercepts; y-intercept $(0,1)$. (c) Vertex $(-2,0)$; x-intercept $(-2,0)$; y-intercept $(0,4)$.

(3) Determine from the graph by using vertex form $y=a(x-h)^2+k$ and one additional point.

Practice Exercises 13: Polynomial Functions

(1) (a) Not polynomial, variable in denominator. (b) Polynomial, degree 1. (c) Polynomial, degree 4. (d) Not polynomial, contains fractional exponent $x^{1/2}$.

(2) (a) $f(x)=x(x+3)(x-4)$. (b) $f(x)=(x+3)(x+1)(x-2)(x-5)$. (c) $f(x)=(x+2)^2(x-4)$.

(3) (a) zeros -4 mult. 1 crosses, -3 mult. 3 crosses; end behavior like $4x^4$. (b) zero $\frac{1}{3}$ mult. 2 touches, zero 1 mult. 3 crosses; end behavior like x^5. (c) zeros 0, $\pm\sqrt{3}$, all mult. 1, all cross; end behavior like $4x^3$.

(4) (a) $f=(x+1)(x+4)(x-3)$: x-intercepts $-4,-1,3$; $y=-12$; crosses at all; like x^3; max turning points 2; above $(-4,-1)\cup(3,\infty)$, below $(-\infty,-4)\cup(-1,3)$.

(5) (b) $g=6x^3(x+4)$: x-intercepts $-4,0$; $y=0$; crosses at both; like $6x^4$; max turning points 3; above $(-\infty,-4)\cup(0,\infty)$, below $(-4,0)$.

(6) (c) $h=x-x^3=x(1-x)(1+x)$: x-intercepts $-1,0,1$; $y=0$; crosses at all; like $-x^3$; max turning points 2; above $(-\infty,-1)\cup(0,1)$, below $(-1,0)\cup(1,\infty)$.

Practice Exercises 14: Rational Functions and Asymptotes

(1) Domains: (a) $x\neq-3$. (b) $x\neq 2,-4$. (c) $x\neq-2$.

(2) Asymptotes: (a) VA $x=\pm2$; HA $y=2$; no oblique. (b) VA $x=1$; no HA; oblique $y=\frac{1}{3}x+\frac{1}{3}$. (c) VA $x=-1,0$; hole at $x=1$; HA $y=0$; no oblique. (d) no VA; no HA; oblique/polynomial asymptote $y=x^2+x+2$.

(3) Graph keys: (a) $F(x)=\frac{1}{x-1}+2$: VA $x=1$, HA $y=2$. (b) $F(x)=\frac{-1}{(x+2)^2}$: VA $x=-2$, HA $y=0$, range $(-\infty,0)$.

(4) Problem I: HA $C=0$; concentration approaches 0 as t increases. Maximum occurs at $t=\frac{1}{\sqrt{2}}$ minutes.

Practice Exercises 15: Exponential Functions and Equations

(1) (a) $x=-4$. (b) $x=\pm3$ if read as $3^{x^2-7}=27^{2x}$; verify original notation. (c) $x=-6$. (d) $x=3$. (e) $x=\frac{1}{2}$. (f) $x=\frac{8}{3}$.

(2) Domains for all: $\mathbb{R}$. Ranges/asymptotes: (a) range $(0,\infty)$, HA $y=0$. (b) range $(2,\infty)$, HA $y=2$. (c) range $(-\infty,4)$, HA $y=4$. (d) range $(3,\infty)$, HA $y=3$.

(3) For $g(x)=4^x+2$: domain $\mathbb{R}$, range $(2,\infty)$, horizontal asymptote $y=2$.

Practice Exercises 16: Logarithmic Functions and Equations

(1) (a) $x=32$. (b) $x=63$. (c) $x=e^2$.

(2) (a) $3\log x+2\log y$. (b) $4\log x-3\log y$. (c) $3\ln x+2\ln(x+2)$.

(3) (a) $\log(x(x+3))$. (b) $\log\left(\frac{x^3}{y}\right)$. (c) $\ln(x^2(x-1))$.

(4) (a) domain $(0,\infty)$, VA $x=0$, increasing, x-intercept $(1,0)$. (b) domain $(-2,\infty)$, VA $x=-2$, increasing, $(-1,0)$. (c) domain $(3,\infty)$, VA $x=3$, increasing, $(4,0)$.

(5) Sketch using the above transformations/asymptotes/intercepts.

(6) (a) $x=64$. (b) $x=\log_3 7$. (c) $x=\frac{1}{3}$. (d) $x=-2$. (e) $x=-\frac{7}{2}+\frac{\sqrt{85}}{2}$. (f) $x=3$.

Practice Exercises 17: Systems of Linear Equations

(1) $(3, 7)$.
(2) $(2, 4)$.
(3) $(3, -1)$.
(4) $(4, 3)$.
(5) $(\frac{7}{2}, 4)$.
(6) $(\frac{3}{4}, 6)$.
(7) Inconsistent.
(8) Dependent.
(9) Independent.
(10) One solution: $(1, 3)$.
(11) No solution.
(12) Rectangle dimensions: width 12 m, length 18 m.
(13) Numbers: 9 and 17.
(14) 6 pens and 3 folders.
(15) 110 hamburgers and 140 hot dogs.

Index

www.ingramcontent.com/pod-product-compliance
Lightning Source LLC
LaVergne TN
LVHW082333200926
848481LV00012B/3332
* 9 7 9 8 3 1 9 7 3 7 3 7 3 *